SCIENCE, RELIGION, AND THE FUTURE

SCIENCE, RELIGION, AND THE FUTURE

A COURSE OF EIGHT LECTURES

BY

CHARLES E. RAVEN, D.D.

MASTER OF CHRIST'S COLLEGE AND
REGIUS PROFESSOR OF DIVINITY IN THE
UNIVERSITY OF CAMBRIDGE

εἰς οἰκονομίαν τοῦ πληρώματος τῶν καιρῶν, ἀνα-
κεφαλαιώσασθαι τὰ πάντα ἐν τῷ Χριστῷ, τὰ ἐπὶ
τοῖς οὐρανοῖς καὶ τὰ ἐπὶ τῆς γῆς.

Ephesians i, 10.

CAMBRIDGE
AT THE UNIVERSITY PRESS
1943
REPRINTED
1968

CAMBRIDGE UNIVERSITY PRESS
Cambridge, New York, Melbourne, Madrid, Cape Town, Singapore, São Paulo

Cambridge University Press
The Edinburgh Building, Cambridge CB2 8RU, UK

Published in the United States of America by Cambridge University Press, New York

www.cambridge.org
Information on this title: www.cambridge.org/9780521074377

First published 1943
Reprinted 1968
This digitally printed version 2008

A catalogue record for this publication is available from the British Library

Library of Congress Catalogue Card Number: 43–8868

ISBN 978-0-521-07437-7 hardback
ISBN 978-0-521-08170-2 paperback

To

my friends

JOHN HENRY ARTHUR HART, B.D.

sometime Fellow of St John's College, Cambridge

and

HENRY ST JOHN HART, M.A.

Fellow and Dean of Queens' College, Cambridge

'The learned doctor deals out labels to his predecessors in the scientific criticism of the Gospels, as they did to the Jews of long ago. Wheat and tares alike are rooted up and dried and pressed and labelled and discussed—and you learn as much about the living plant as you can learn of a fox by contemplating its mask and pads and brush, each cured and mounted for display as the relics of an animal (vermin or not vermin) worthy to be hunted, or shot, or trapped, according to the custom of the country. For scientific observation of living things the systematist must wait upon the biologist. You must go to the earth and hide and wait, if you want to see the vixen play with her cubs.'

J. H. A. HART, *The Hope of Catholick Judaism*, pp. 114–15.

CONTENTS

PREFACE

The proper purpose of a Preface is to state the standpoint and intention of that which follows. It is an apologia though not necessarily an apology.

In the present case it is enough to say that these lectures were delivered in a time of world-wide war. Plainly the men and movements which direct the thought of mankind had dismally failed. To talk about the triumphant march of science would be as ironical as to sing 'Like a mighty army, moves the church of God'. Both science and religion must take a share of blame for the appalling catastrophes which they ought to have been able to prevent. They represent the most important formative influences in the educational and indeed in the intellectual life of the world; and the result of their efforts in the recent past has been a holocaust unequalled in history. The complacency with which their leading representatives lay the blame upon the social order, or Nazism, or the politicians, or the devil makes it clear that they do not recognise their responsibility—or rather, since we are all involved, that we are all still impenitent and self-satisfied.

It is this conviction which explains my very critical interpretation of the recent history of biological and theological studies. I assume that there has been a cause of the present calamities less superficial than the Chauvinism of the victors of 1918, or the Caesarism of Mussolini, or the evil genius of Hitler. This cause seems to me to consist in the inability of mankind to make sense of his world, to agree upon the significance of existence, and to co-operate for its welfare; and in the consequent appearance of incompatible, indeed of violently contrasted, ideologies. For this the blame must rest upon those who failed to adjust human thought and life to the new knowledge which the past century has disclosed—that is upon the intellectual, moral and religious teachers of mankind.

PREFACE

Having myself begun the study of Genetics with Professor Bateson and of Christian doctrine with Professor Bethune-Baker in the same term in 1907 and having been a teacher (of a sort) all my working life, I can at least see the failure in myself even if I am mistaken in discovering it in others. Somehow the people responsible for education, for shaping and propagating ideas and for developing civilisation have allowed science and religion to become antagonistic with results disastrous to them both and devastating to the life of men.

It is the purpose of the first four of these lectures to indicate the history of that disaster; and of the second four to consider how, if at all, it may be retrieved.

I had prepared a series of appendices on various controversial issues ranging from the pathological theology of to-day to the industrial melanism of Lepidoptera. But these would be too brief for the specialist and too technical for the general reader, and are (rather reluctantly) omitted.

A word in conclusion. These lectures and their subject were arranged by the Divinity Faculty of the University of Cambridge. I would not seek to make the members of that Faculty partners in my guilt: but at least they encouraged me to a task which I should not have undertaken on my own initiative. To them and to the audience who 'stayed the course' so generously with me I owe a great debt.

C. E. R.

February 1943

I

THE 'NEW PHILOSOPHY': SEEING
LIFE WHOLE

Gratulor ergo fortunis meis meritasque Divino numini gratias ago, quod inani illa leptologia, quae Philosophiae titulo nostrâ etiam memoriâ Scholas occupavit, fastiditâ et facessere jussâ, in locum ejus successit solida et experimentis inaedificata Philosophia.

JOHN RAY, *Synopsis Stirpium Britannicarum*, Preface, p. 8

1. INTRODUCTORY

To mention Science and Religion in the same sentence is, as we are all aware, to affirm an antithesis and suggest a conflict. The two words, like Oxford and Cambridge or even God and the Devil, are charged with memories—memories in this case of a bitter struggle for freedom from theological trammels, of arrogant claims arrogantly answered, of sanctities wantonly profaned and, more recently, of timid and not quite honest overtures for a truce. Most of us if we are over thirty-five have ranged ourselves in one camp or in the other; and if we are younger are inclined to say, 'a plague on both your houses: between you you've made a bloody mess of mankind: personal relationships and no isms or ologies for us—if we survive the result of your mistakes'. All of which is surely a pity: for with every respect to W. S. Gilbert it is simply not true that

> Ev'ry boy and ev'ry gal
> That's born into the world alive
> Is either a little Liberal
> Or else a little Conservative

and the difference between the Liberal and the Conservative is scarcely more arbitrary than that between the scientific and the religious. The Latin comedian is in this matter far more truthful than the English.

THE 'NEW PHILOSOPHY': SEEING LIFE WHOLE

We are all human, and nothing human is or should be outside our interest: Terence tells the truth. We are all human: and as such it is our business and our joy to make sense (whatever sense we can) of our world, and to enlist all our faculties for that task. To say, 'I am a scientist; truth is my concern; and therefore I refuse to admit the importance of anything that cannot be weighed and measured', is as silly (yes, silly) as to say, 'I am a Christian; God is my business; and therefore anything that is not in the Bible (or St Thomas or Luther) is irrelevant'. Yet both are still constantly said and almost universally implied by people otherwise not mentally defective. That no one ever acts consistently upon either of these prejudices only makes the resulting confusion more confounded.

Of course, if scientists like to say, 'We are not interested in the whole of experience, but in a small and specialised aspect of it; and in order to avoid entangling ourselves with religion we will call the study of dead organisms biology and the study of mechanical reflexes psychology', they are at liberty to do so, provided they tell us frankly that they make no claim to cover the characteristically human activities of life and are content to be the gadget-mongers of a society whose control must then rest with persons of larger outlook.[1] Similarly Christians are not forbidden by creed or canon to set nature and grace in contrast, to divide history into sacred and profane, to reject Copernicus and Darwin and revert to the world of magic and make-believe (which is, alas, what many of our modern 'revolutionaries' are doing): but if they do so, let them admit that they have abandoned the teaching of Jesus and an incarnational philosophy and the over-hard task of making sense of the twentieth century.

But man, whether you call him *Homo sapiens* or *Homo faber* or *L'homme qui rit* or a Worshipping Animal, man who cannot but appreciate the world, and cross-question it, because, after all, he wants to live in it more abundantly, will not be deeply impressed

[1] For a valuable warning against confusing science with technology cf. J. R. Baker, *The Scientific Life*.

by the restrictions and evasions of either party. He will go on insisting that he is one man not three men, artist, scientist and saint; that for fullness of development he must play all three parts; and that life is the happy exercise of all the appropriate activities. He is the same man in his garden when he considers the lilies and wonders how best to stop his Madonnas from turning brown, and in his cathedral when he partakes of the Sacrament and wonders how he can get over his dislike of the celebrant's intonation; and perhaps he is as much a disciple of Jesus in the one setting as in the other. He is the same man, even if he lives in many places and relationships and at different levels of his being; and the whole business of effective living for himself and for society depends upon his making himself consistent and 'all of a piece'.

For if he is a normal person he will know that the joy with which a sunset touch or the Hallelujah Chorus affects him is not far removed from that with which he discovers the solution of an intellectual problem[1] or finds a thrill of satisfaction in the doing of honest work. He will have felt the moment of self-abasement in which hell opens for him and he is stripped of his illusions and his pride; he will have felt not less surely the moment of exaltation when he is caught up into a heaven of harmony and permanence; and he will know that, paradoxical as the contrast appears, there can be no finality in it since he, in hell or heaven, is still one and the same. Indeed, he will welcome such tension, not because in the shallow modern jargon tension is itself creative (it is *not*; love, not strife, alone creates), but because the point of tension is always, if he strives faithfully to resolve it, the point of growth.

In consequence he will also know that the method by which these paradoxes of experience are to be resolved must not be

---

[1] J. R. Baker, *The Scientific Life*, pp. 11–23, has a chapter on the 'act of discovery', but seems to regard it as peculiar to scientists. J. Y. Simpson has collected many examples of the 'creative moment', cf. *Landmarks in the Struggle between Science and Religion*, pp. 61–73. There is a full treatment of it in Dr Rosamond Harding, *An Anatomy of Inspiration*.

sharply divided into categories and labelled 'scientific approach' or 'religious outlook'. Whatever the problem, physical or metaphysical, biological or theological, he will approach it in the same way, collecting and studying the relevant evidence, testing and experimenting, classifying and interpreting, striving like Lucretius to 'seek out the causes of things' and asking like Clerk Maxwell 'what's the go of it?' Of course the data will vary: those necessary for an investigation in eugenics will differ from those in archaeology; but the principles of inductive research are the same in every field and even the experiences of mystics and saints are amenable to their application.

This is a point which has not been made sufficiently clear; and in religion is not yet perhaps universally acknowledged. Scientists still occasionally talk about the scientific method as if it were their own exclusive and strictly patented possession; and Christians less occasionally hanker after infallibilities and, opposing faith to sight, identify it with credulity. But for most of us the realm of the irrational, of magic and make-believe, can no longer co-exist with the realm which it is our business and joy to explore and understand and interpret. Our universe is no longer a duality of reason and superstition, or of nature and grace, or of secular and sacred. Varied as it is, profound as are the differences between its aesthetic, intellectual and moral aspects, we whose concern is with knowledge of it will employ the common-sense technique of observation and experiment, the familiar *Novum Organum* which in three hundred years has won its way to universal usage. We may adopt different standards of verification and test our results by different criteria: that is a matter for enquiry and discussion. But even if our application of the technique is defective, induction remains our instrument: whatever our subject, we are in that respect 'all scientists now'.

2. THE CONDITIONS OF THE TASK

There are of course three phases in the use of this method and these, because superficially they vary according to the subject studied, have led men to suppose that artist, thinker and saint possess peculiar faculties, esoteric sources of knowledge and specialised means of applying it—whence the rubbish that is still talked about the artist's intuition or the saint's religious instinct. There are three phases of study, each demanding its own aptitude; all normal men possess capacity for the three activities, but differ in the extent to which one or other is adequately developed.

A. EYES AND NO EYES

The first is awareness and appreciation. It is to perceive and become fully sensitive to the object of our interest. Mr Henry Williamson, himself a fine observer of nature, writes of Richard Jefferies that he had supreme powers of vision; and he means not only acute physical sight but the ability to differentiate what he saw, the imagination to appreciate it, the memory to retain the impression of it. 'Eyes and no eyes' is a caption appropriate to spiritual and historical as well as to artistic or scientific vision. We may use different assistants to supplement our natural capacity: one does not apply a microscope to the stars or a spectroscope to the Bible; but the basic activity which observes and records is the same whatever field of study occupies us, whether our awareness is more properly called sight or faith.

It is a truism that even in the matter of naïve perception folk' differ enormously in the speed, subtlety and range of their faculties. Sight, hearing, smell, taste, touch—men vary in them almost as much as in psychic, intellectual and religious sensitiveness. Some, whether from the perfection of their physical apparatus or from native ability to receive and register impressions, have such gifts as make their fellows seem blind or deaf. This is of course a matter in which training, and especially the training in infancy of whose technique we are so lamentably ignorant, can do much. My Dutch friend who from the massed choir of birds in the dawn chorus of mid-May can pick out each individual performer and name it and whistle its song has brought a naturally delicate ear to perfection by years of attention. My botanical colleague in his golfing days and in spite of keeping his eye on the ball has brought back *Potentilla verna* from the links on the Gogs and *Orobanche picridis* from Worlington. John Ray described his friend Francis Willughby as having a brilliant flair for the discrimination of specific distinctness; and J. H. Fabre, who was praised by Darwin as 'that inimitable observer', by patient watchfulness opened up a wonderland for the naturalist and the philosopher in his *Souvenirs entomologiques*. It is proper to suppose that in their own fields the great artists and musicians, poets and prophets and mystics have sensibilities which open to them ranges of experiences of which the rest of us are scarcely conscious. To deny the truth of their vision because we do not ourselves share it, is to put the telescope to a blind eye and declare that there is nothing to be seen.

Naïve perception, before it becomes sophisticated, seems to be normally and often disconcertingly reliable. The 'instinctive' reaction of a child with its uncanny skill in detecting pose; the immediate impression at a first meeting which if we can recover it proves to be exact; the freshness of vision which enables some men to look at a familiar object as if they had never seen it before and so to discover novelty where the rest of us are merely bored; the sensitiveness to human relationships which serves as a sixth sense in estimating quality and enabling sympathy; these things, and I would add the spiritual awareness which belongs to the life of prayer, indicate that our unspoiled aptitudes can be trusted far more than we commonly suppose.

But plainly at a very early stage interpretation and conscious selection begin to influence perception—and to influence it both for better and for worse. Man's awareness is never that of a photographic plate, automatic and indiscriminate. From the first, even if he actually sees all that is visible, he only perceives what in some degree attracts his attention; and what he perceives is already coloured by association and explained by experience. We see what we expect to see; and our actual consciousness of it is inescapably subjective. In these days, when the recognition of our relativity has encouraged every kind of extravagance, it is obvious that, in the arts at least, subjectivity has often wilfully repudiated all external references. Looking at the work of some recent painters, normal folk must often echo the dictum ascribed to Adolf Hitler: 'If you see things like this, you are mad; if you only pretend to see them, you are bad: in either case Himmler will take charge of you.' If we do not want to hand them over to the Gestapo, at least we would like to see them dealt with as Macaulay dealt with Robert Montgomery. Unfortunately in these days of strain theology is as much infested with the eccentric, the untrained, as are the arts. For it is assumed that anyone from a tennis champion to an ex-Communist can without further qualification instruct the world on Christian doctrine and ethics; and bishops are as easily found as they were in the days of Guy Thorne to give benedictions and forewords to books which by another Communion would unhesitatingly be put upon the Index. This is no doubt mostly the theologians' own fault—they have made their subject so technical that the normal man prefers eccentricity to dullness. But if truth is not to suffer, it is important that naïve perception should be reinforced by sound knowledge and not left to the unchecked vagaries of individual prejudices and temperaments.

Anyone who has had experience of the requirements of exact observation will know how profoundly vision itself is influenced by wishes and expectations. The trained ornithologist, for example, hears constantly stories of bird behaviour which his friends profess to have seen but which plainly reflect the psychology of the *Jungle Book* or the *School of the Woods*; and he knows from his own experience how easily sentimental and anthropomorphic ideas distort not only the interpretation but the actual content of what is seen. Strict objectivity, the sort of exactitude which science demands, is not common even in the perception of nature: it becomes more difficult with history; and perhaps hardest of all in religion. Take, for example, the revelations which purport to be given through automatic scripts or trance or spiritualism: you will hardly find one that is not wholly subjective, a fair sample of the tolerably educated, liberal-minded, middle-class culture of a decade or two ago, hardly one that shows a spark of originality or unexpectedness. To compare them with the New Testament, with the stories of a Messiah whose character was in flat contrast to popular expectation and whose Jewish contemporaries were the last people on earth to accept His claims, is to see the difference between spurious and authentic observation.

B. ANALYSIS: ITS POWER AND ITS PERILS

Next to perception is therefore the process of sifting, classifying, examining and interpreting the data: indeed, the curiosity which initiates this process is inextricably combined, as we have indicated, with perception itself. We see what intrigues us, we pigeonhole what we see, and like a dog with a buried bone we disinter and gnaw it in moments of leisure, extracting what savour we can from it, comparing it with similar specimens in our experience and trying to discover the 'how' and the 'why' of its existence. To classify each particular event under its appropriate heading, to investigate the interplay of the parts within the whole with a view to formulating explanatory hypotheses or laws, to arrange the resulting generalisations into a system, this is roughly the business of the second or intellectual stage of our study. It represents our scientific rather than our aesthetic activity: but its method is similar whether we are dealing with problems of the laboratory, or with the abstractions of pure mathematics, or with the interpretation of historical events, or with the philosophy of religion.

So much is, or should be, platitude. Man may not be sapient: he is at least inquisitive, and delights in cross-questioning his environment.

But in this business of analysis he is exposed to dangers which he seldom fully realises and never entirely escapes. Having split up his object of study—be it water into hydrogen and oxygen or Genesis into J E and P—he assumes first that these sources exactly represent that from which they are derived, and secondly that they are the only stuff that matters. The first assumption involves

8

a denial of the fact of wholeness—the fact that any 'whole' by mere virtue of uniting within itself these analysable factors possesses that which defies analysis. The second denies the fact of novelty—the fact that no one contemplating hydrogen and oxygen separately could recognise them as water; it denies that the parts in union behave differently and are different from the same parts in isolation. A whole in spite of Euclid is greater than the sum of its parts; for in the act of analysis its integrity is destroyed.

It was regrettable though perhaps inevitable that in the earliest days of the scientific movement, when the method of induction was beginning to challenge scholasticism, this fallacy of analysis was generally ignored. When Descartes on the basis of his vivisections and analogies argued that although man was something more than a machine all animals were automata,[1] he was refuted rather by the hearts than by the brains of his critics—though John Ray argued that a creature which could calculate the breadth of a stream and wait for its master at a cross-roads could not be wholly devoid of intelligence. But when the triumphs of the Newtonian physics were followed by two centuries of successes gained by confining attention to mechanistic categories, the Cartesian absurdity was applied not only to dogs but to their owners. Indeed, it was actually asserted that since what could not be weighed or measured did not exist, thought was solely a series of laryngeal movements—a fantasy still maintained even when its authors admitted that these movements had neither weight nor measurement! And Dr Hogben tells us that 'most psychologists are behaviourists nowadays'![2]

In the exigencies of conflict with the tradition scientists, as we shall later describe in detail, were forced into a position in which they consented to confine their attention to mechanistic and

---

[1] It must be admitted that it was and still is the belief of the Roman Catholic Church that 'brutes are as *things* in our regard...we do right in using them unsparingly'. J. Rickaby, *Moral Philosophy*, p. 248.

[2] *Science for the Citizen*, p. 1047: I do not know on what evidence this dictum is based.

9

quantitative categories: hence science has come to mean not the intellectual treatment of the *totum scibile* but only such knowledge as can be established by weight and measurement. But this is a temporary, arbitrary and eminently unsatisfactory restriction, as may be seen by the controversy arising as to the scientific status of psychology and as to the claim of history to be a science. The 'New Philosophy' was the title given in the seventeenth century to the inductive method and the whole range of sciences resulting from it; and when the Royal Society was founded, though religion and politics were barred, this was solely for practical reasons and did not prevent industrial, economic, speculative and metaphysical problems being freely discussed.

The damage that has been done to man's intellectual welfare by this over-rigid differentiation between 'science' in the narrow sense and other subjects of enquiry and research can hardly be over-estimated. It has led to a glorifying of departmental and specialised studies and to consequent rivalries and snobbishnesses, to the anatomising of life, the consequent destruction of mental and social integrity, and the disappearance of integrating and unifying principles alike in philosophy and in religion. Unfortunately, owing to its long and uncritical acceptance, this sectionalism has become a habit; the scientist who insists that he cannot keep his science and his citizenship in separate compartments, the historian who sees his work as a continuation of that of the biologist studying evolution, the Christian who refuses to agree that his religion is an amiable idiosyncrasy, these are exposed to almost universal misunderstanding if not to rebuke. 'Cobbler stick to your last.'

This exaggerated and, as all must hope, temporary estrangement is a thing of recent growth, and quite foreign to the true relationship between science and theology. At its best periods and in its best representatives, in the third century with Origen, in the thirteenth with Thomas Aquinas, in the seventeenth with the Cambridge Platonists, there has been a steady insistence upon the 'congruity' of nature and grace, upon the need 'to see life steadily and see it whole', and upon the authority of reason. It

THE 'NEW PHILOSOPHY': SEEING LIFE WHOLE

was as a result of one of the most rigorous investigations ever under-taken—an enquiry which lasted for three hundred years and was pursued by a succession of outstanding scholars—that the Christian system of doctrine was first formulated. The learning and disci-pline of the great Scholastics developed this system into a com-plete and coherent philosophy in terms of the best thought and science of the time. Even when the new philosophy, following upon the Renaissance and adopting the inductive method, opened up a fresh and immensely enlarged knowledge of nature, there was at first little sign of serious conflict. Copernicus was never persecuted; Stensen, the first modern geologist, became a Catholic bishop; Ray, the pioneer of modern botanical and zoological studies, was proud to say 'Divinity is my profession'; Boyle and Newton spent more energy upon theological enquiries than upon chemistry and physics. To us with a century of antagonism in mind and realising, as it is easy to do after the event, how incompatible was the new with the old *Weltanschauung* it may be difficult to believe that controversy was not immediate and that neither side felt the incongruity of the other.[1] But in Britain at least there was no open breach until a century ago, and there has never been wanting a steady supply of scholars who have been both scientists and theologians—and that without insincerity or conscious ob-scurantism in either field.

C. LIFE AND LIFE ABUNDANT

The fact is of course that though there may occasionally be those who deserve the epitaph 'This man decided not to live but know' they are perhaps fortunately rare. Knowledge divorced from life, knowledge specialised, anatomised, desiccated, speedily becomes valueless and is always dangerous. No doubt the man who collects

[1] Books like A. D. White, *Warfare of Science with Theology* and J. W. Draper, *Conflict between Science and Religion* can hardly claim to be im-partial, and in fact read back the struggle of last century into earlier times.

11

tram tickets like Professor Hardy's mathematician[1] (from whom he is perhaps not far removed) can comfort himself by the thought that his activities, if useless to any except himself and the small circle of his fellow-collectors, are at least blameless and inoffensive. But mere expertness without the check of practical application, mere intellection in the abstract is too far removed from life to be satisfying or fruitful; for it is from experience as we live it out that the mind must constantly derive the provision of its raw material, the testing of its progress, the verification of its results. The thinker who takes his feet from the ground will not long enjoy keeping his head in the air. The third phase of our task, beyond perception, and interpretation, is the living out of what we have seen and known; and here neither the senses nor the mind alone but the whole self is involved.

In these days when Professor Eddington has brought home to us the extreme queerness of the material world as disclosed to the new physicist we are all conscious of the truth of the ancient Greek adage that 'eyes and ears are bad witnesses'. If a series of complex equations is the only medium capable of expressing modern concepts of space-time, we hardly need to be reminded that 'words like nature half reveal and half conceal the soul within'. We are in fact aware that intellect proceeding by the mechanism of the spoken thought is a very clumsy instrument for expressing experience; that in order to express it we have to transpose it into a different key or rather degrade it on to a lower mode of being; and that even so the available symbols fail to do exact justice to any except the most elementary events.

No doubt the abstractions of the mathematician can be defined precisely; but they are at the opposite pole of our experience to life. Chemistry and physics, the inorganic world, can be described with tolerable accuracy, and for fear of losing this accuracy scientists are unwilling to extend their studies to the realm of the animate. The study of mankind if it is to be exact requires a medium and a technique capable of expressing personality and

[1] *A Mathematician's Apology*, pp. 59, 81.

personal relationships: language, as anyone who has ever tried to write to the newly bereaved must know, is insufficient even when handled with the skill of a poet and the sympathy of a friend: only immediate and living contact, the touch of life upon life, can transmit a true report.

This is what lies behind the best of the modern protests against tradition in art and thought. When Picasso attempts to depict the impression which an object makes he may or may not succeed, and the result may, to anyone else, be hideous or disgusting: but at least he is trying to get away from the objectivity of the coloured photograph to the presentation and interpretation of human feeling. When we accept Kierkegaard's demands for 'existential' thinking, we are similarly insisting that philosophy is not mere cleverness and logic-chopping but must learn to express life in its tensions and its simplicities, and that religion is not a system of beliefs but a personal relatedness. Painter and prophet are saying in effect that man is alive and must not behave as if he were a camera or a calculating machine; that he is alive and that this fact should influence his perception and his intellect. Contemplation and dialectic must lead on to action if full life is to be attained.[1] This is also what is involved in the Christian belief in the Word made flesh.

Scientist, historian and theologian are all men concerned with description and interpretation—the two being hardly separable, since no description can be entirely free from selective and therefore explanatory influences. Such men's business is to discover categories which can properly explain to other human beings the significance of existence as man experiences it. To describe in terms of mechanism and the purely physical may be accurate enough so far as it goes; it is easy, objective and useful in enabling man's control of things; it is inadequate in dealing with life and intolerable when applied to human relationships. To introduce organic categories at once takes the task outside the realm of

---

[1] Cf. the brilliant exposition of this necessity in Bergson, *Les Deux Sources*, pp. 236–49 (English trans. pp. 181–99).

weight and measurement and involves the need for more sensitive methods of investigating the living object than we at present possess; we have to make the best of what we have; plainly imagination and sympathy, unnecessary for the interpretation of the star or the stone, become essential for the explanation of life. But the categories employed by biology cannot do justice to personality or the strictly personal ranges of human èxperience; and while this is left out of account no complete scientific system is attainable. It is only at the personal level, by personal analogies, that we can hope to describe and interpret the whole. Mechanistic and even organic concepts involve so large a measure of abstraction and arbitrary delimitation that the pictures which they present cannot but be partial and distorted. Personality itself, the actual living person, would seem to be the only medium competent to express and explain to persons their universe of experience.

The objection commonly taken to this argument (and it is of course the Christian doctrine of Incarnation) is against its obvious anthropomorphism; and in these days of relativity it is doubly difficult to feel that the objection is justified. When an artist paints a picture and calls it a Stag at bay or (in the modern mode) a Gazelle we assume in the first case at least that he is being naturalistic and objective. When he formulates a Law of Gravity or a theory of the Origin of Species we assume (or assumed until recently) that he is describing not what he imagined to be real, but reality itself. On the face of it there seems no reason to suppose that man's art or man's science will give a less biased and subjective interpretation than that which can be found in a man's whole self. If we can find truth in the aesthetic and intellectual activities of humanity, how much more in the whole character or selfhood of which those activities are a part? No doubt our descriptions will be more exact or factually correct as they get nearer to the bleak abstractions of mathematics. But as an interpretation of the universe nothing is so unsatisfying as a series of pointer-readings. To be exact at the expense of leaving out everything that matters is only justifiable if we insist upon supplementing such descriptions by the use of other and more concrete methods; and the more these do justice to the intimacies and profundities of personality the more will they convince us of their adequacy. That is why great drama, whether on the stage or in real life, is the most illuminating, initiatory and effective of all human means of expression. The *Prometheus Vinctus* or the *Oedipus Tyrannus*, the myths of Plato and the mysteries of Eleusis, convey a deeper as well as a more generally appreciated philosophy than the lecture room or the pulpit. Incidentally they point forward to the supreme mysterium or revelation which Paul the Apostle proclaimed when he preached Christ crucified and risen, and to the dramatic rite in whose action Christians have found the clearest symbol and instrument of their faith.

To express in the quality of our life what we have perceived and interpreted will be to achieve a measure of integration; and, in proportion as we have seen and understood rightly and are loyal to our experience, we shall attain adjustment to our environment and a high survival value. At its best such expression of reality deserves Aristotle's description of man's true end—to live immortally—for it is that communion with God and co-operation with His will which has been the constant goal of religion: and to attain it would be to embody in our own selves the experience of fulfilment and wholeness with which most of us are only familiar at second-hand or in rare moments of exaltation.

But if the attainment is beyond our reach, it is nevertheless the goal to which the finest human endeavour has been devoted. Seers, thinkers and saints have accepted it as their objective: even when tragically imperfect or wilfully perverted it has still had unequalled power to liberate and inspire and unite mankind: pre-eminently it has influenced the course of history and survived the downfall of civilisations. To dismiss it as illusory or irrelevant to the study of reality or the welfare of mankind is to be deliberately blind to the most characteristic and significant facts of human evolution. Science can no more dispense with religion than religion can dispense with science. Only the exercise of all man's powers in their due place and proportion is worthy of man's task and opportunity.

3. THE WORTH OF THE TASK

Upon the Church the promotion of such exercise should be a primary obligation. In the first century of its life, when the spreading of the good news outside the Jewish world was only just beginning, the purpose of the missionaries was declared by their greatest leader. He was not content merely with the evangelising of mankind from east to west, or even with the welding of humanity into an organic society. In his moments of clearest insight he saw the whole creative purpose, the whole universe of our experience finding its consummation and unity in the one Christ who represents the sum of the values embodied in the creation and is himself the incarnation not only of those values but of the eternal God from whom they are derived.[1] Christendom too soon renounced this essentially Christian appreciation of nature and history—though before the mistake was made the great Greek apologists and above all the great Alexandrians, Clement and Origen, had given to it noble expression in the richest and most satisfying scheme of thought that the Church has ever known. If we in these days take up the task of interpreting all experience in the light of Christ, we shall only be returning to the work which St Paul and St John undertook and which in the Church's greatest days was orthodoxy.

In so doing we shall gain new and rich resources of spiritual life. It is the peculiar genius of the prophets and psalmists of Israel who were the forerunners of Jesus that they rejoiced in the works of the Lord and filled their thoughts of God with the imagery of starry heavens, of mountains and woodlands, of streams in the desert and valleys thick with corn. In all the tragedy of their history, in defeat and captivity, they never doubted that the earth was the Lord's and that he was righteous. Jesus inherited

[1] Cf. the verse quoted on the title page, Ephes. i, 10, and the first two chapters of the Epistle to the Colossians.

and revealed this knowledge of the sacramental value of nature and drew from its simple rhythm material for his deepest lessons. No one can study the pioneers of the scientific movement without realising that they were men who found in the 'new philosophy' and in the observation and interpretation of the natural order a religious experience. There is in all their work a sense of curiosity, of wonder as of children exploring a new country, an enthusiasm and a depth of feeling which, while it does not blind them to the difficulties or the magnitude of their adventure, sends them to it with certainty of its worth and confidence in its goal. About the temper of De l'Ecluse or Malpighi, Boyle or Ray there is the authentic quality of spiritual greatness; and unless scientific studies become professional or narrowly abstract, and therefore dehumanising, their devotees keep that quality still. In it they are at one with the great poets of our race, with Wordsworth in his nature mysticism and Shelley in his scientific insight and Browning in his incarnational philosophy; and at one with the multitude of simple folk who for the past century increasingly have found aesthetic, intellectual and spiritual release in the study of some branch of natural history, and a thrill of wonder at the discoveries that each decade has achieved.

II

THE AGE OF TRANSITION: THE CHILDHOOD OF SCIENCE

Insanum quiddam esset, et in se contrarium, existimare ea quae adhuc numquam facta sunt fieri posse, nisi per modos adhuc numquam tentatos.

FRANCIS BACON, *Novum Organum*, Aphorism VI

1. CHRISTIANITY OLD AND NEW

THE exhortation to man to use all his powers for the interpretation of truth and the attainment of the good life sounds attractive even if platitudinous. The insistence that health involves wholeness and that wholeness must include and indeed be centred upon religion is also neither unreasonable nor unfamiliar. In these days of totalitarianism we are all of us accustomed to the claim that the individual can only fulfil himself as he brings his whole life into subordination to the collective purpose; we have seen enough of the dynamic or demonic energies which such integration can produce; and we admit that an ideal incarnate in a man, be it Communism in Lenin or Nazism in Hitler, gains an unprecedented and stupendous influence. Moreover, if it were a choice between a Nazi or Communist and a Christian ideology as the inspiration of such totality, Mr T. S. Eliot is surely right in believing that we in this country at least would prefer the Christian as more in keeping with our past and less depressing for our future.[1] Let so much be granted.

Unfortunately the prospect of such a development seems not merely remote but under present circumstances frankly unattainable. It is necessary to speak plainly on the matter. While the Christian religion as professed by the churches still clings restrictively to a *Weltanschauung* that is demonstrably unscientific, to superstitions that violate the intelligence and to conduct that

[1] Cf. *The Idea of a Christian Society*, p. 13.

18

shocks the morality of modern man, no such consistency as is essential can be expected. To be a Christian or at least to hold official position in the churches is to accept formulae parts of which can only be explained by being explained away or else to keep secular knowledge and religious belief in permanent estrangement. To ask men to live in two such irreconcilable worlds is to imperil the possibility of the wholeness of life which is our need. Christendom is still and apparently contentedly wedded to a system of thought and to some extent also of ethics and organisation which is pre-scientific, indeed almost medieval. It has not yet completed its move out of the habitation elaborately constructed for it in the thirteenth century.

The majesty and the logicality of this system are indeed so impressive as to keep in thrall to it, even to-day, a large section of Christians. It was wrought out, upon the basis of the long and splendid labour of the great Christian scholars of the first five centuries, by the almost equally great Scholastics of the thirteenth. From the standpoint of the art and science, the politics and the religion of that day it was a thing complete, so lucid that only trivialities seemed to be left for enquiry, so rational that to accept its premises was to be committed to its conclusions. It provided for the whole life of man as then understood and organised: it possessed enormous prestige and resources so that rebels were helpless and explorers like Roger Bacon were endangered; and even when with the triumph of Nominalism blind obedience to authority took the place of intellectual effort and when with the Reformation the system of papal government to which it seemed committed was successfully challenged its general ideas and outlook survived and were taken over by the Lutherans and reaffirmed in an almost more rigid form by the Calvinists.[1] It is the broad outline of this system which still claims an almost exclusive

---

[1] For an excellent account of the general beliefs and practices of the medieval Christian cf. B. L. Manning, *The People's Faith in the time of Wyclif*. This shows how limited and in the main superficial were the changes effected in the fifteenth and sixteenth centuries.

right to call itself Christianity. To understand the present position, especially as between science and religion, that system in its plain and popular form must be described.

It was based upon an acceptance of the books of the Old and New Testaments as the Word of God, irreformable and inerrant, and consequently upon the cosmology declared and implied in them and especially in Genesis or Revelation. These books had long been officially or traditionally interpreted: but the Church had condemned the great Origen who had adjusted their crude literalisms to a reasonable philosophy by the use of allegory, and had also condemned his great successor and rival Theodore who had laid down the principles of a sound historical criticism. Treated as oracular, but in practice heavily expurgated, they had been elaborated into a dogmatic schema to which Aristotelian science and Augustinian theology were the other main contributors.

Under this system, altered in detail but unchanged in principle by the Reformation, Christians believed vividly in a personal God conceived in anthropomorphic terms and often identified with Jesus: they also believed not less vividly in a personal devil. For them as for the medievals or indeed for the pagans of the Augustan age the air was full of angels and fiends: if Shakespeare did not quite believe in a real Oberon or Ariel, he certainly believed in real witches and probably in a real Setebos. He also believed in a local heaven 'above the bright blue sky' and a local hell within or beneath the earth. This earth, now the battleground between celestial and infernal powers, had been created by God at a fixed date about four thousand years before the Christian era.[1] In it man had been placed in a state of brief but perfect bliss until God's intention had been thwarted by the successful guile of Satan, who thus got all the secular world into his power. To counteract his growing control and as a last remedy against the

[1] Though commonly associated with Archbishop Ussher, this dating is very much earlier than his time.

ruin of the divine purpose God had sent his own Son to earth, and there in the appropriate theatre and at the appointed time the drama of his birth, death and resurrection was staged and played. God's legitimate anger with rebellious mankind was appeased by the obedience and self-sacrifice of his Son; a divine Corporation, the Church, was established; and those who accepted the prohibitions and duties enjoined by it were secured if not in this life at least in the life to come. The world of nature and history save as the environment of the Church—the sea on which the sacred ark voyaged!—was in the last resort irrelevant. Only the supernatural life of the Church gave it any meaning. In itself it was fuel of fire and would one day be consumed. To that day, the day of wrath, the day of the Second Coming of the Son of God, the attention of the Christian was emphatically directed. Then God whom Jesus had declared to be the loving Father would change his character, reverting to the ferocity and tyranny of his Old Testament habits: even Jesus himself who had once prayed 'Father, forgive them' would now lay aside mercy and deal out justice and vengeance: sinners for whom he had formerly sought as a shepherd for his lost sheep would now be flung into hell and there would suffer endless flames and torments—a spectacle to increase the beatitude of heaven.[1]

How this strange medley of fact and fable, of truth and falsehood, of good and evil had gained its influence can only be appreciated by a detailed study of the slow stages by which the message of the Apostolic Church was devitalised and distorted. It happened in the course of acclimatising Christianity to the environment of an age morally and intellectually exhausted, economically and politically bankrupt, and for our present purpose need not be described.[2] But of the extent of the change there can be no question. The system, which had shaped the whole

[1] On this point Aquinas, *Summa*, S.Q. xcv, 3 and Jonathan Edwards, *Works* (ed. 1817), vii, p. 480, agree: cf. E. Westermarck, *Christianity and Morals*, p. 52.

[2] I have dealt with it in detail in my book *The Gospel and the Church*.

ordering of European society, creating not only a vast organisation for its own propagation but vast vested interests in the social, economic and political life, continued almost unchallenged until the seventeenth century; and even now persists and occasionally comes into power. Some of the strangest features of it are still accredited parts of Christian orthodoxy.[1]

But its significance for our present purpose is less in its statements than in the outlook upon nature and history which accompanied it. Though there was an underlying trust in the reliability of reason derived from the long succession of Greek and Christian philosophers, there was no body of organised and verified knowledge. The universe was governed arbitrarily by the will and act of a God whose ways were past finding out: in it there was no concept of law or even of probability: crazy things might happen at any moment, and miracle or magic would always explain them. Modern man, whatever his views of the supernatural, will recognise that the cure of paralysis by the removal of a guilt obsession is an event on a totally different level of probability from an act of levitation. Yet to the man of Shakespeare's time it was certain that the devil manifested himself in the form of a black boy, that witches could transport themselves on wings if not on broomsticks, and that frogs and lemmings were spontaneously generated in the clouds. Instances could be multiplied indefinitely to prove not only that there was no appreciation of what could and could not happen, but that except in respect of the knowledge guaranteed by ancient tradition all progress was vitiated by the universal acceptance of the arbitrary and the chaotic. Until the middle of the sixteenth century knowledge was derived either from the Scriptural and ecclesiastical tradition or from the Latin classics, Cicero for philosophy and ethics, Pliny for natural history, Ptolemy for astronomy, Dioscorides for physic, and the records of the alchemists reveal the curiosity of the resultant practice.

[1] A charmingly presented version of it was published in 1942 in Mr C. S. Lewis' *Broadcast Talks.*

THE CHILDHOOD OF SCIENCE

How long these superstitions persisted can be seen by a couple of illustrations relevant to the University of Cambridge. In 1645, when John Nidd of Trinity first began his researches in biology and was observing the breeding of frogs in a glass vivarium, a woman was hanged in the town for keeping a tame frog which was sworn to be her imp;[1] and in 1669, seven years after the foundation of the Royal Society and eight years after Isaac Newton came into residence, the University entertained Cosimo de' Medici with a dissertation denouncing the Copernican astronomy.[2] Very significant are the three divisions of Henry More's *Antidote against Atheism* published in 1654 and the work of one of the ablest and best men of the time. In the first book we have a brilliant and original presentation of the Christian Platonism which More shared with Whichcote, Cudworth, John Smith and others—a philosophy as profound as any that England has produced. In the second is a painstaking but curiously uneven account of the ordering of nature as giving evidence of design and benevolent purpose. In the third, which More himself seems to regard as the most convincing proof of theism, are werewolves, poltergeists, the Pied Piper of Hamelin and all the witchcraft and magic which Joseph Glanvill and More afterwards re-stated in that strange volume *Saducismus triumphatus*. And if anyone suggests that More was a Christian and not a scientist, he may fairly be reminded that More was a close disciple and correspondent of Descartes and an original member of the Royal Society: and moreover that the whole contemporary world of scientists and theologians, believers or sceptics, were agreed in their acceptance of the 'novity' of the world, of creation as a sudden fact, and of the fiery cataclysm which would bring the story to a close. Indeed, as Dr Trevelyan puts it, 'All that minority of mankind who thought seriously about life was genuinely Christian'[3]—at least to the extent of sharing the contemporary and characteristic Christian outlook.

[1] Cf. Ray, *Wisdom of God*, 2nd ed., pt. II, p. 84, and Cooper, *Annals*, III, p. 398.

[2] Cf. Cooper, *Annals*, III, p. 536. [3] *England under the Stuarts*, p. 60.

It is however a complete mistake to suppose that because this traditional *Weltanschauung* was Christian, therefore the advocates of the 'new philosophy' were themselves unchristian or were persecuted by Christians. The idea of a conflict between religion and science in the seventeenth century is based upon the fact that among the thousands who were done to death by the Inquisition or the lawcourts for offences against established ethics or religion a few like Servetus and Bruno had been scientists—as well as queer and difficult people. But in fact the evidence for such a conflict is slight, and has been exaggerated in the interests of later controversy. In reality, as we have seen, the early scientists, whether in Europe generally or in this country, were Christians and in many cases clergy; they proclaimed that the study of nature was in itself a religious duty; and they challenged the old system of belief and education because it was concerned with dry-as-dust conventionalities instead of with the manifold and fascinating works of the living God.

In this they were at first hardly conscious of making large innovations and certainly had no desire to pose as rebels or destroyers. Rather they claimed to take up the age-old task of the Church by interpreting afresh the meaning of God's creation, appealing for sanction to the great utterances of psalmists and prophets and to the explicit injunctions of Jesus. Christianity had always professed a concern with truth; at its best it had fostered a lively interest in nature; the new studies based upon observation and experiment would supplement the old; since new and old alike were devoted to a fuller understanding of God, there could be no serious clash between them. So the 'new philosophers' went ahead with confidence, accepting the records of Scripture, the findings of tradition and the evidence of research as alike data for the formulation of a true system.

To the modern student this is apt to be bewildering. When Lyell,[1] commenting upon the history of geology, expresses his surprise that John Ray can quote Genesis and Cicero's *De Finibus*,

[1] Cf. *Principles of Geology*, I, p. 46.

24

Stenson and Robert Hooke side by side, he is reflecting back into the seventeenth century the controversies of the nineteenth. To do so is in fact as absurd as it is for Mr J. G. Crowther to ridicule Clerk Maxwell for his Victorian feudalism;[1] or as it would be for me to censure Mr Crowther for his out-at-elbow Marxism. For the business of appreciating the ideas of the past is hard enough even for the historian, and in dealing with times of transition becomes doubly difficult. But if we are to understand the origin and influence of the scientific movement, we must recognise first that it owed an immense debt to the Platonist and Aristotelian traditions which had given to Europe a respect for reason and a belief that the world of the sense-perceptions bore an intelligible relationship to the world of general principles; secondly, that the Christian schema, though plainly defective in its details as to the origin and the end of creation, yet had established a belief in an order and movement in the course of events which, if expressed in too crude a teleology, yet gave scientists their impulse towards the discovery of evolutionary developments; and thirdly, that Christians themselves, dissatisfied with their own traditions and with the friction and maladjustments of secular affairs, were the pioneers in what they regarded as a definitely religious adventure. Men turned to the study of science not as rebels against Christianity but as dedicating themselves to a new and more plainly Christian crusade.

2. THE PIONEERS OF THE NEW PHILOSOPHY

The forerunners were born in the fifteenth century, Columbus before 1450; Leonardo in 1452; Erasmus in ?1467; Copernicus in 1473; Paracelsus in ?1493. Biologists came a little later: Fuchs in 1501 and other early botanists soon after; Vesalius, the anatomist, in 1514; Gesner, the first great naturalist, in 1516. By the end of the century medical schools and dissection, physic gardens and herbals, collections and pandects were becoming part of the life of the great Universities at Bologna and Padua, at Paris and Mont-

[1] Cf. *British Scientists of the Nineteenth Century*, II, pp. 348–9.

pellier, at Basle and Heidelberg, Belon and Rondelet, Jean and Gaspard Bauhin, De l'Obel and De l'Ecluse were lecturing, travelling, writing. Men were beginning to be aware of the world of plant and animal life around them.

So far there had been little recognition of the wider significance of the new studies, and with the possible exceptions of John Caius, William Gilbert and Thomas Mouffet hardly a single Englishman of much importance. But it was to Francis Bacon rather than to Girolamo Cardano or to Pierre de la Ramée or Giordano Bruno that the origin of the 'new philosophy' must be ascribed. His wide range of interests, restless intellect and legal training enabled him to take all knowledge for his province, a new intellectual world for his aim, and induction or the building-up of conclusions upon verified evidence as his method. That he planned far more than he fulfilled, failing through lack of concentration and through personal ambition, and endangering the influence of his project by the scandal of his fall, must be recognised: but the *Instauratio* gave to the world a clear and comprehensive outline of its new tasks and to his own countrymen a summons which, if it met with no immediate response, evoked by the middle of the century a movement at once conscious of its intentions and effective in its achievements. England started late in the field of science: but thanks to Bacon it started well. He died childless: the Royal Society was in a very real sense his offspring.

To say so is not to disparage the significance of notable visitors like Comenius, or notable exiles like Samuel Hartlib and Henry Oldenburg; nor of natives who owed nothing to Bacon, William Harvey who learnt his physiology at Padua, or Joseph Mead the first of the Cambridge Platonists and his two great successors at Christ's College, John Milton and Henry More; nor of the members of the Invisible College who gathered at Gresham College in London and Wadham in Oxford, Robert Boyle, 'father of chemistry and brother of the Earl of Cork'; John Wallis, the first great mathematician, of Cambridge and Oxford; John Wilkins, son-in-law of Cromwell and stepfather-in-law of

Archbishop Tillotson, of Oxford and Cambridge; and Robert Hooke, the first professional scientist, with his twisted body and indomitable brain, his cantankerous temper and his queer attractiveness, who did much of the work for which Boyle and others took the credit and who could never resist provoking the snubs which he never ceased to resent.

In Britain the opportunity arose for giving birth and self-consciousness to the stirring life of which the age had been so long pregnant. Bacon had heralded the travail; the pangs were the religious and political conflicts of Civil War and Commonwealth. In them it would be an exaggeration to say that the old order died: but very certainly the horrible records of savagery and persecution which in the Thirty Years' War made abortive the intellectual life of Europe were represented in Britain by the milder distresses of the twenty years in which the King was killed and the Commonwealth enthroned and deposed, and in which a brilliant generation nourished upon exaltations and agonies turned away from the squabbles of the sectaries and the intrigues of the politicians to the systematic exploration of the new world which telescope and microscope were revealing. It is in the work of John Ray, Francis Willughby, Nehemiah Grew, of Richard Lower, John Mayow, and Thomas Sydenham; of Robert Hooke, Edward Lhuyd and John Woodward; of Isaac Newton, John Flamsteed and Edmund Halley that the symbolic and epoch-making struggle of the Stuart period has its worthiest outcome.

These men drew to them and to England the admiring attention of the older centres of science. Marcello Malpighi, greatest of contemporary Italians, sent his *Anatome Plantarum* in manuscript to London, and saw it published there in 1675 and 1679. Anton Leeuwenhoek reported to the Royal Society in 1677 his discovery of spermatozoa, though it did not appear in their *Transactions* till 1683. Tentzel, the German, wrote to them in 1697 about the mammoth bones found in Thuringia which his fellow-countrymen declared to be a freak of nature but he recognised as of an elephant. Tournefort, king's professor of botany in Paris, sent a personal envoy in 1698 to see Sloane's Jamaican plants and to express by a special visit to Black Notley his homage to the genius of Ray. Leibnitz in 1705 paid Newton the unwelcome compliment of a wholly fictitious accusation of plagiarism from him in regard to the method of fluxions. We had started late: but we made up for lost time.

The fact is that the country was, for the sixty years following the Commonwealth, singularly rich in men of scientific ability and that the Royal Society with its regular meetings, its very remarkable first Secretary, and its excellent periodical gave a range and unity to their work. Newton in Physics and Ray in Botany and Zoology were pioneers of outstanding merit, but they were accompanied not only by the remarkable men already named but by a large number of others. By the end of the century great progress had been made in mathematics, in astronomy and physics and to a less degree in chemistry; comparative anatomy was established and physiology was beginning to be studied; the flora and fauna of Britain and Western Europe had been classified and described; a good start had been made with geology and physical geography.

Moreover, and this for our purpose is equally important, a serious contribution had been given towards changing the general outlook in philosophy and religion. The medieval educational system—'the scragged and thorny lectures of monkish and miserable sophistry'[1] as Milton unkindly called it—had been shaken from its supremacy. Bentley had set up a laboratory for the Italian chemist Vigani and an observatory on the great gate of Trinity for Cotes; he had previously introduced the principles of the higher criticism in his Dissertation on Phalaris and had expounded natural theology in his Boyle lectures. The scientific study of languages and particularly those necessary for archaeology, and of antiquities and history, had developed greatly since the days of Camden and the chroniclers. If the results were chiefly valuable in recovering knowledge of early times and literature and were

[1] *The Reason of Church-government urged against Prelaty.*

largely used to explain and support tradition, yet a just appreciation of the past is a primary necessity for its wise amendment. Even the fixed scheme of thought, the ideas of the 'novity' of the earth, of creation as a specific act, of the deluge, and of the end of the world, ideas which Thomas Hobbes the sceptic or John Locke the philosopher held as firmly as Newton, were beginning to be questioned—and hitherto without raising serious protest. By the beginning of the eighteenth century it looked as if the traditional Christian *Weltanschauung* was being changed without controversy— as if the new philosophy was going rapidly forward to the religious and scientific achievements which Boyle and Ray had prophesied. Another half-century similar to that from 1650 till 1700 might have seen the swing from old to new accomplished.

Unfortunately the level of effort was not sustained; the great age came to an end; and after 1720 there seemed to be little or no first-rate ability available. A moral, intellectual and religious rot had set in.

3. THE TENSION BETWEEN SCIENCE AND RELIGION

To enquire·into the reasons which led to this collapse would take us too far from our subject—though it is worth observing first that the Church, partly as a result of its losses by the Puritan emigrations, the Act of Uniformity and the non-juring Schism, partly owing to its failure to take advantage of the new philosophy, was scandalously unfit to uphold religion and hardly attempted to do so, it was either high and dry or soft and low, in either case moribund, and secondly that science, having been to some degree an escape for men sickened of religious and political controversies, had neither the influence nor the will to check the abuses, the profligacy, drunkenness and wild gambling of the times. As Wesley's work demonstrated, only a Christianity at once cognisant of the best contemporary thought and passionately proclaiming a real and appropriate message could meet the need.

But though the general corruption was a principal cause for the slowing-down of the scientific movement, a more special reason was the fact which the eighteenth century demonstrated that until more radical research in chemistry and physics had been carried out, the biological sciences could hardly advance further. So long, for example, as combustion and the heat of the body were ascribed to an element, fire or phlogiston, neither chemistry nor physiology could get far; so long as the flood was made responsible for all the major changes in the earth's surface, no appreciation of stratification or of the age and sequence of fossils could be expected. It was not solely the prestige and presidency of Newton that made the Royal Society ignore natural history; for Linnaeus and his school, great as were their labours in taxonomy and nomenclature, added no more than their British contemporaries to the larger issues of science. It was inevitable that there should be a pause until the work of Lavoisier and others liberated chemistry from the influence of its ancestral alchemy, and the growth of observation made hypotheses of magic and witchcraft unreasonable. Then progress could go forward again.

Meanwhile, and all through the eighteenth century, in Britain there was a steady succession of good field naturalists, Pennant, Barrington, Gilbert White and others, who kept up the interest in and enthusiasm for nature; most of these were men of strongly Christian conviction; several were ministers in the Anglican or other communions. There was no conflict or organised opposition to science on the part of the churches; and the work of adjusting their theology and religious activity to scientifically verified results went on, slowly indeed (for interest was not widespread) and haltingly (for men of genius were rare), but without much tension or friction. Butler, the ablest and best bishop of the century, wrote his *Analogy*; Wesley, in training his evangelists and class leaders, gave them a simpler version of Ray's *Wisdom of God in the Works of Creation*; Paley, besides his *Evidences of Christianity*, produced his only less influential *Natural Theology* which Charles Darwin declared that in his youth he knew almost by heart. In spite of Tom Paine there was no conflict between science and religion in Britain until the middle of the nineteenth century.

Very typical of the persistence of teleological and mediating ideas was the series of Bridgewater Treatises endowed in 1829 by the eccentric eighth Earl and selected by the President of the Royal Society—who in selecting asked the help of the Archbishop of Canterbury and the Bishop of London. Of the eight volumes that by Thomas Chalmers repeated the arguments from design in a form rather different from those employed by Ray and Paley; those by William Prout and by William Whewell developed these arguments with regard to physiology and to the appropriateness of the environment; and that by William Buckland sketched the progress of geological studies and exposed itself to violent attacks from the champions of the strict litteralness of Genesis. Only that by William Kirby, the oldest of the team and a pioneer in entomology, repeated the Miltonic picture of creation—the seas suddenly prolific and the teeming of the womb of earth; ridiculed Lamarck's evolutionary concept; and appealed to Charles Lyell, the rising geologist, in support of the fixity of species. The volumes, though very uneven in merit, fairly represent the scientific outlook of the time; and to an intelligent reader it must have been apparent that, though relations between tradition and research were strained, the leaders on both sides (as is evident from the letters of the period) thought it possible and eminently proper to preserve a compromise at all costs.[1]

Still more significant—and more disturbing!—was the famous eirenicon by Philip Henry Gosse, the popular naturalist of the sea-shore, whose reputation has been alternately whitened and blackened by his son's two biographical studies.[2] Gosse combined unusually narrow views on religion—he was a Plymouth brother until he found the sect too lax—with a real enthusiasm for the study of nature. After a number of popular books he published

[1] For a recent and generous recognition of the value of the Treatises cf. F. Wood Jones, *Design and Purpose*, pp. 39–42, 57.
[2] *Life of P. H. Gosse* (1890) and *Father and Son* (1907).

31

in 1857 an ambitious 'attempt to untie the geological knot' which he named *Omphalos*. This was an answer to the venerable conundrum: 'Which came first, the hen or the egg?' It assumed that all life was in a sense continuous, a sort of circle; and that God at the creation set it off at a particular point, creating trees with rings of growth and rocks with fossiliferous strata and Adam with a navel, although all these vestiges of the past were at the moment meaningless since the trees had not grown nor man been born. Apparently the effect of this argument did not dawn upon its author who, poor man, was so confident of having solved his problem that he printed a huge edition of his book at his own expense. But he got a letter from Charles Kingsley in which were the sentences: 'Assuming the act of absolute creation—which I have always assumed as fully as you—shall I tell you the truth? It is best. Your book is the first that ever made me doubt it, and I fear it will make hundreds do so. Your book tends to prove this— that if we accept the fact of absolute creation God becomes *Deus quidam deceptor*.... It is not my reason but my conscience which revolts here.'[1]

It is something of an irony that it should have been Gosse who prepared the way for Darwin.

[1] *Life of P. H. Gosse*, pp. 280–1.

III

THE CONFLICT: A STORM IN A VICTORIAN TEA-CUP

They find now that they have got rid of an interfering God—a master-magician as I call it—they have to choose between the absolute empire of accident and a living, immanent, ever-working God.

CHARLES KINGSLEY, *Life*, II, p. 171

[NOTE. A short explanation is necessary for those who will criticise this chapter as disproportionately long and needlessly detailed. For such it is (1) the record of a momentous event—one of the seeming trivialities which had a far greater influence than the wars and laws, the kings and popes that fill our history books; (2) an illustration of a philosophy of history: for history (though I should not subscribe to Carlyle's definition of it as the story of great men) is not and cannot truthfully be made a diagram of impersonal and economic forces: its formative events are not primarily Cleopatra's nose but Caesar's (and Cleopatra's) queer, but alas! largely unknown, reactions to it; still less are they the ownership of the means of production, as Communist propaganda asserts: history is primarily a study of human character and motive.]

1. THE OPPOSITION TO DARWIN

THE geological conflicts of the thirties and forties of the nineteenth century might, in spite of the fury of Dean Cockburn and the believers in the literal inerrancy of Genesis, have passed away without serious consequences. There is a stage in childhood when the youngster's awkward questions about sex and digestion can be answered by fables and no harm be done. On this matter psychologists who inveigh against fairy tales and demand lectures on physiology in the nursery may be safely written down as 'prigs, pedants and prose-mongers'. Genesis, the six days and the deluge, would have taken their proper place in the folklore and poetry of the race, and religion and science might not have been fatally antagonised, had it not

33

been for the much more embittered controversy over Darwin's *Origin of Species*. It may well be that historians in the future will ascribe two 'decisive battles of the world' to the sixth decade of the nineteenth century, and put alongside the great lock-out of 1852, which drove the A.S.E. and with it the whole Labour Movement from a policy of co-operative production to a policy of class war, that other great conflict which took place on Saturday, 30 June 1860, at the meeting of the British Association in Oxford, and made overt and abiding the cleavage between traditional Christianity and progressive science.

The story of that conflict, comparatively recent though it is, can perhaps hardly yet be told fully and fairly. The main facts of it are of course familiar—the story has been recounted perpetually: but details are not easy to discover; for the protagonists were all great men in the Victorian sense of those words and therefore their remains are embalmed in the monumental 'Lives' whose purpose was not history but eulogy, and whose evidence, though less blatantly expressed than that of Hanoverian epitaphs, is hardly less partial and misleading. *De mortuis nil nisi bonum* is not a precept which makes for truthful records. The reaction against them, despite Mr Lytton Strachey, has not yet made the historian's task much easier; for in the main it has fastened upon and exaggerated matters of secondary importance.

Moreover, in the growth of scientific opinion and temper the Darwinian controversy took place at a turning-point. At that particular juncture things were said which would have been impossible, indeed almost meaningless, ten years earlier or ten years later. One of the chief protagonists, writing of his great rival in 1894, said: 'The thing that strikes me most is how he and I and all the things we fought about belong to antiquity: it is almost impertinent to trouble the modern world with such antiquarian business.'[1] Those who watch the growth of children will know how from time to time points are reached in which the whole outlook and habit seem to change almost overnight: our play-

[1] *Life of T. H. Huxley*, II, p. 373.

fellow of yesterday is a stranger; the old jokes, the old games, have lost their charm; we find instead of the chuckle of glee a slightly supercilious wonder as to what the old fool thinks he is playing at. So it was with science in Darwin's decade: a growing-point had been reached.

The fact is that the outbreak at the British Association was an explosion almost fortuitous in its incidence—the sort of thing which the wiser scientists and theologians had been for some time trying to avert, and which if it had not come at that exact psychological moment would have been insignificant. As it was, it cleared the air. After it the stifling sense of oppression caused by a tradition, which few accepted yet none overtly renounced, was dissipated: men could breathe and speak more freely. But like all explosions it also destroyed; the damage was not so obvious at the time as it has become since; but it was even then evident and far-reaching.

The explosion was fortuitous. The actual *casus belli* and the personnel of the combatants gave the discussion its importance; but the causes of the strife and the temper displayed were derived from events which had little to do with Darwin or indeed with the issue between science and religion. Like many other cataclysmic happenings its origins are to be sought in fields remote from the actual issue.

That Darwin himself expected controversy not with theologians or the religious but with his fellow-scientists is clear from the mass of evidence in his *Life* and elsewhere. Except for a postscript in a letter to Lyell, 'Would you advise me to tell Murray' (his publisher) 'that my book is not more unorthodox than the subject makes inevitable',[1] there is no anticipation of religious opposition; and in fact Canon Tristram was the first zoologist to publish his acceptance of natural selection;[2] Charles Kingsley was steadfast in his support;[3] and even after the quarrel in 1860 J. R. Green wrote: 'I was much struck with the fair and unprejudiced way in

[1] Letter of March 1859, *Life of C. D.* II, p. 152.
[2] *Ibid.* p. 205 note.   [3] E.g. *Ibid.* p. 236.

which the black coats and white cravats of Oxford discussed the question.'[1]

On the other hand, from scientists he expected, and received, fierce criticism. Thus, writing in 1857 to Asa Gray, he records: 'The last time I saw my dear old friend Falconer' (Hugh Falconer, the palaeontologist) 'he attacked me most vigorously. . . . "You will do more harm than any ten naturalists will do good. I can see that you have already *corrupted* and half-spoiled Hooker".'[2] 'I have accustomed myself. . . to expect opposition and even contempt,'[3] he said to Hooker in October 1858, and 'Owen will bitterly oppose us,'[4] to Wallace in August 1859. Indeed, his own manner almost invited opposition; for his modesty made him magnify the revolutionary character of his hypothesis, and his single-mindedness ('It is an accursed evil to a man to become so absorbed in any subject as I am in mine'[5]) drove him to worry almost intolerably about the reception of his book.

Here is in fact a circumstance of much importance for understanding the controversy. Charles Darwin was a man of almost unbelievable naïveté, with a childlike and therefore lovable conviction of the supreme importance of his own concerns and with a seriousness towards his own books such as few men have given even to the Scriptures. That he forfeited his appreciation of music and of natural beauty, his religion and indeed all his other interests save his family affections to the one obsessing pursuit, is not surprising in view of the extent to which his work dominates all his correspondence. Living under terribly restricting conditions a life of more than monastic discipline he came near to fostering a persecution fantasy, and to ascribing to scientists in general a hostility to himself and a doctrine of creation, both of which he grossly exaggerated. This led him to be highly provocative. When he wrote, 'All the most eminent palaeontologists, namely, Cuvier, Owen, Agassiz, Barrande, Falconer, E. Forbes, etc., and all our greatest geologists, as Lyell, Murchison, Sedgwick, etc., have

[1] *Life*, II, p. 323. [2] *Ibid.* p. 121. [3] *Ibid.* p. 138.
[4] *Ibid.* p. 162. [5] *Ibid.* p. 139.

unanimously, often vehemently, maintained the immutability of species',[1] he deserved the retort that he 'betrays an ignorance or indifference to the matured thoughts and expressions of some of those eminent authorities';[2] and when he further insisted that all who do not agree with transmutationist ideas 'really believe that at innumerable periods of the earth's history certain elemental atoms suddenly flashed into living tissues'[3] he may have been logical but was certainly neither tactful nor accurate. Producing such irritation he might fairly be said to have asked for the kicks which he anticipated.

If his ill-health and consequent isolation were responsible for this misjudgment of the situation, his social position was (there can be little doubt) a further source of trouble. Ever since the foundation of the Royal Society there had been something of a cleavage between the virtuosi, the nobility and gentry who patronised and were entertained by scientific experiments and the working scientists who made a livelihood by supplying them.[4] The distinction persisted into the middle of the nineteenth century when a large number of the House of Peers whose education effectually prevented them from gaining any knowledge of science and who actually knew nothing beyond their stud farms and game preserves could write F.R.S. after their names from having filled the presidential chair. In the eighteenth century the aristocratic amateur was accepted as an ornament of an otherwise vulgar pursuit: in the nineteenth science had proved its importance, and the professionals were no longer in need of patrons. The controversy about the presidency of the Royal Society is only one symptom of a widespread resentment against the pretensions of men of rank and leisure. When the Edinburgh Reviewer of the *Origin of Species* begins his essay by describing Darwin in the words

[1] *Origin of Species*, ed. I, p. 310.
[2] *Edinburgh Review*, cxi, No. 226, p. 501.
[3] *Origin of Species*, p. 483.
[4] A symbolic example is seen in Westminster Abbey where Newton's monument at the entrance to the choir is balanced by one to James Lord Stanhope!

'Of independent means he has full command of his time for the prosecution of original research: his tastes have led him to devote himself to Natural History: and those who enjoy his friendship are aware...'[1] he is giving expression to his own scorn of the dilettante and the amateur. We shall not do justice to the critics of Darwin unless we remember how hard was the life of the working scientist in days when endowments were few, when even medical men were despised as socially inferior, and when too often the praise and rewards went to those whose money or birth gave publicity to their work.

Whatever the cause it is clear that, as Huxley afterwards put it, 'the supporters of Mr Darwin's views in 1860 were numerically extremely insignificant...if a general council of the church scientific had been held we should have been condemned by an overwhelming majority'.[2] But the bitterness of the struggle was not only or chiefly due to the disparity in numbers. Personal issues beyond those already mentioned played a large part.

The real antagonist to the new doctrine was, as Darwin had foreseen, Richard Owen. Not only was the great anatomist a man who had had to struggle from obscurity through the vagaries of a difficult medical apprenticeship to a position of security, but having become the protégé of the Prince Consort his labours in zoology and palaeontology had gained a well-earned fame. In consequence not only did he account himself the proper leader and spokesman of all British scientists, but his nature, never very amiable, became almost intolerably jealous and egoistic. The official biography, the second volume of which is largely devoted to social functions and lectures to the Royal Family, naturally gives no idea of the extent either of his cantankerousness or of his isolation.[3] But his own writings, especially his unsigned reviews, are sufficient proof. No man can ever have used anonymity so

[1] *Edinburgh Review, l.c.* p. 488.
[2] *Life of C. D.* II, p. 186.
[3] The *D.N.B.* does something to repair the omission; but much more is due.

blatantly as a means of self-praise:[1] the reviews are in fact largely selections from 'Professor Owen's' acknowledged works to prove how vastly superior they are to those which are being criticised. When it is realised that (as Darwin delicately indicated) he was prolix and obscure in his own utterances and also quite incapable of consistently sustaining an argument, the danger of his pre-eminence in science will be recognised. His authority was usually asserted by *ipse dixit*: if not, he could always quote some cryptic sentence from one of his own books or lectures to prove priority over his rivals or to show how clearly he had disposed of their results.

Unfortunately for Darwin the publication of his work intervened upon an existing and embittered quarrel between Owen and his former pupil, Thomas Henry Huxley. Huxley assigns the cause of this to Owen's arrogance in styling himself Professor of Palaeontology (and therefore Huxley's chief) when he was delivering a course of lectures in Huxley's College, the School of Mines, in 1857.[2] 'An internecine feud rages between Owen and myself (more's the pity)',[3] he wrote in March 1858. Shortly afterwards, in his Croonian Lecture on 'The Theory of the Vertebrate Skull', he had attacked the speculation of Oken and thereby, as he claims, 'demolished the superstructure raised by its chief supporter, Owen, "archetype" and all';[4] and of a subsequent meeting he writes: 'Shall I have a row with the great O. there? What a capital title that is they give him of the *British* Cuvier. He stands in exactly the same relation to the French as British brandy to cognac.'[5] When such was his temper Owen was not the man to sit down under his insubordination or to lose any opportunity of punishing it. Darwin might write, indirectly pleading for generosity—'If my views are *in the main* correct, whatever value they may possess in pushing on science will now depend on the verdict pronounced by men eminent in science'[6]—but Owen

---

[1] Cf. *Life of C. D.* II, p. 304.   [2] *Life of T. H. H.* I, p. 142.
[3] *Ibid.* p. 158.   [4] *Ibid.* p. 141.
[5] *Ibid.* p. 161.   [6] *Life of R. O.* II, p. 92.

knew that Huxley was a supporter of the Darwin-Wallace hypothesis, and could hardly be expected to be unbiased. Despite the
letter, Darwin's fears were fulfilled and, though his authorship of
it was never acknowledged, Owen delivered his condemnation in
the April number of the *Edinburgh Review*. If less violent than that
which Adam Sedgwick the geologist had delivered in the previous
month in the *Spectator*, it is of course very much more complete
and very much more deadly.

Its line has been partially indicated. Darwin is a distinguished
amateur: he has done several promising bits of work: in this
present book there are three or four observations about the habits
of ants which are contributions to knowledge. But that is all.
When we ask what is the evidence on which the grandiose claims
made for Natural Selection are based we find literally nothing
but guesswork and speculation. The theory which we have been
asked to await so eagerly is merely a rehash of Lamarck with no
more substance than the theories of Buffon or of the *Vestiges*. To
suggest that if we do not accept it we must believe in a succession
of miracles is ridiculous: we do not accept the views which are
here imputed to us, but are not so arrogant as to dogmatise without
evidence upon 'the continuous operation of the ordained becoming
of living things'. Then after a series of arguments against details
of Darwin's contentions he concludes with a paragraph which, if
his words mean anything, endorses Linnaeus' doctrine of the
immutability of species. Some of his detailed criticisms are sound
—though his ridicule is clumsy and his scorn often cruel. Some
of his protests are justifiable—Darwin did assume, through
modesty, an exaggerated detachment from his contemporaries.
But the review is a malicious piece of work; and it is hardly
surprising that neither he[1] nor his biographer admitted authorship
of it.

[1] In *Life of R. O.* II, p. 96, is a letter from Sedgwick, 'Do you know the
author of the article in the Edinburgh?' No answer is given.

A STORM IN A VICTORIAN TEA-CUP

2. CHURCHMAN v. SCIENTIST

So far, though Sedgwick had charged Darwin with materialism and mentioned the word 'atheism', there had been no specific attempt to make the issue one between science and religion. If it had remained a mere quarrel between biologists like that which later split them between Poulton and Bateson, little serious damage would have been done. It was left to Samuel Wilberforce to enter the lists against Darwin as the champion of Christendom against the infidel; and by his intervention do enormous harm.

Samuel Wilberforce, third son of William the hero of negro emancipation whom Cobbett declared to be the bitterest enemy of the white slaves of mine and factory, was a churchman of a not uncommon type. Ambition concealing itself under a genuine concern for the institution convinced him that for its sake and by God's will he was destined to be its leader and that it was his religious duty to work for that end. He was already regarded as the most conspicuous of the younger bishops, though he had lost reputation and the favour of the Court by his equivocal handling of the Hampden controversy and by the secession of his brother and brother-in-law to Rome. If he was to recover his position it was important that he should come out as the spokesman of orthodoxy against a cause which was weak and unpopular. Darwin's book gave him his opportunity.

He was not in fact ill-qualified for the task. The man who could explain away his nickname of 'Soapy Sam' by saying 'You see I am always in hot water and always come out with clean hands' was no fool and had a real gift for debate. He had got his First in Mathematics at Oxford and so could claim familiarity with science and scientists.[1] He knew that Owen, the foremost living biologist in Britain, had denounced Darwin and regarded him as a pretentious amateur. He was a regular contributor to the *Quarterly* and could review the *Origin of Species* there at length. Whether he was encouraged to do so or helped in the doing by

[1] Cf. W. C. D. Dampier, *A History of Science*, p. 299.

41

Owen is open to question. Darwin and Huxley both regarded his speech at Oxford, a speech delivered after the composition but before the publication of his review,[1] as 'crammed' and spoke of Owen as the crammer. That he had seen the *Edinburgh* of April before writing his criticism is certain; and he refers freely to certain pages in Owen's *Classification of the Mammalia*.[2] But the review contains three or four blunders (to which Huxley draws attention) so gross that Owen can hardly have vetted it and passed them; and there does not seem to be any clear evidence of Owen's hand in it. Probably Owen knew even before the meeting at Oxford that Wilberforce was entering the lists. Apparently during the meeting he paid a visit to the bishop at Cuddesdon. The speech, which was in the main a reproduction of the review,[3] may well have owed something to Owen's company—perhaps even the famous question to Huxley.[4] If so, in suggesting it Owen over-reached himself. In any case Wilberforce, confident of an easy victory, found that he had caught a Tartar.

Circumstances conspired for his undoing. He was on his native ground—the bishop in his cathedral city—the scholar in his own University. A hugely crowded meeting, bored by a long, dull paper by an American visitor, stimulated by several violent but ineffective comments, enthusiastic when at last the familiar and famous orator rose to address them; a subject on which he had written a lengthy article and had the support not only of a large majority of his audience but of the most eminent authorities; an opposition consisting of but a handful, and led by a man young enough to be his son[5] and as yet hardly known to fame, a man neither of Oxford nor of Cambridge, a man whom he could surely

[1] Wilberforce writes of having completed it on 20 May, *Life of Bp W.* II, p. 450: it appeared in July, No. 215, pp. 225–64.

[2] Pp. 98, 100, 103, and later pp. 58, 59.

[3] Cf. *Life of C. D.* II, p. 321; *Life of T. H. H.* I, p. 183.

[4] The kinship of man with the apes had been the issue between Owen and Huxley on 28 June and does not arise directly in the *Origin of Species*.

[5] They were in fact curiously alike, so that many at first sight assumed that they were related.

be permitted to banter. 'He spoke for full half an hour with inimitable spirit, emptiness and unfairness.... He ridiculed Darwin badly, and Huxley savagely, but all in such dulcet tones....' Finally, according to Lyell, he 'asked whether Huxley was related by his grandfather's or grandmother's side to an ape'.[1] And Huxley smote his thigh and murmured to his neighbour, 'The Lord hath delivered him into my hand'.

Thomas Henry Huxley, not alone of the Victorians, was a great man whose stature increases with distance. Starting life with few advantages of birth, 'kicked into the world a boy without guide or training or with worse than none', he plunged into vice and nearly made shipwreck. Arrested by the influence of Carlyle's *Sartor Resartus*, finding 'a resting-place independent of authority and tradition' in science, given 'a view of the sanctity of human nature and a deep sense of responsibility' by falling in love,[2] he gained from the struggles of his youth something of the passion for truth, of the fearlessness and prophetic fervour, of an Epictetus or an Augustine. With a better brain than Matthew Arnold and more human kindness than Leslie Stephen he shared their high gifts of integrity, moral earnestness and devotion to human welfare, combining in a very unusual degree energy with sensitiveness. He was hardly a great biologist, for he had none of that enthusiasm for nature which delights in the study and observation of the living organism, and took no interest in field-work until he retired and began to grow gentians at Eastbourne.[3] But as 'a sort of mechanical engineer'[4] he was admirably equipped for problems of form and function, and in the sciences of the laboratory and dissecting-room was a match for Owen or any other.

He had a quick and incisive wit and had trained himself to a readiness of speech which, if less eloquent than Wilberforce's, was

---

[1] Cf. *Life of C. D.* II, pp. 321–2: accounts of the meeting differ; the fullest is *Life of T. H. H.* I, pp. 181–8.

[2] The evidence is in his letter to Kingsley, one of the few profoundly revealing documents in Victorian biographies (*Life of T. H. H.* I, p. 220).

[3] *Ibid.* II, p. 443. [4] *Ibid.* I, p. 7.

much more piquant to an excited and slightly cloyed audience. What he actually said has apparently been lost beyond recall: it was one of those utterances whose sheer sincerity and cogency speak direct to the hearer, so that the mere words are forgotten. But its impact was tremendous. Wilberforce was left without defence: Hooker marched in and consolidated the position: the debate was a triumph for the new theory.

3. THE TRIUMPH OF DARWINISM

Yet none could have foreseen how significant it had been. Why was Darwinism so speedily and generally accepted? Why could Kingsley, apparently in 1863, write to F. D. Maurice: 'The state of the scientific mind is most curious; Darwin is conquering everywhere'?[1] Why were Darwin's fears and Huxley's 'beak and claws' so soon proved unnecessary? The answer is not far to seek.

The time was ripe for an explicit disavowal of the belief that creation was either an act once for all 'in the beginning', or a perpetually repeated miracle, 'elemental atoms flashing into living tissues'. The best minds in the Christian Church had perhaps never believed either of them. Certainly in the seventeenth century Ralph Cudworth had insisted and John Ray had agreed that creation was in some sense a continuous process; Redi had demonstrated and Ray had amplified the proofs that spontaneous generation did not happen. Genesis was no doubt taken literally by some: Gosse's ingenious hypothesis is evidence of their existence. But the Linnaean definition 'there are as many species as the different forms which the Infinite Being created in the beginning'[2] had never been universally accepted; and the development of exploration, revealing a vast multitude of living and fossil forms, had made it manifestly absurd. Men were waiting for a natural explanation of the transmutation of species—convinced that in

[1] *Life of C. K.* II, p. 171.
[2] *Genera Plantarum*, p. ii: it must never be forgotten that Linnaeus was a systematist rather than a student of living organisms or a scientific thinker.

some way evolution had taken place, but hesitating to say so while Lamarck and the *Vestiges of Creation* were the only explanations of its method. Darwin supplied a gravely needed want.

The theory which he formulated, though based upon a wide diversity of evidence and expounded with an almost bewildering wealth of detail, was in itself simple, easily demonstrated and eminently congenial to the ideas of the period. Every family photograph, such as was then becoming universal, testified to the fact of variation and suggested if it did not prove that variation was transmitted by inheritance: if young John had his father's hands it was natural to expect that he would make an even finer painter or pianist. To select on the basis of these variations was a practice familiar to everyone who kept pets or stock: a scientist who paid heed to prize bullocks or pedigree fantails was sure of a wide and sympathetic hearing. Biology as practised by an Owen was concerned with long anatomical dissertations or the identification of strange prehistoric beasts. Darwin brought it within the range of the common man.

Moreover, his chief contention was not only intelligible, it had been for sixty years a commonplace of political controversy. Malthus' *Essay*,[1] which Darwin and Wallace both acknowledged as a primary source of their enquiries, had been the bulwark of the orthodox defenders of *laissez-faire*, the basis of the economics of Ricardo, the pretext for every act of exploitation and profiteering. The well-to-do classes had already accepted the 'survival of the fittest' as their own justification. To discover that it was the universal law of nature, responsible for all progress and operating with ruthlessness for the ultimate good of the race, was to find their prosperity given a halo of sanctity. No wonder that the first unpleasant reminder of their animal ancestry was forgiven, and that they applauded one whose work seemed to make material success synonymous with biological excellence, and to demonstrate that interfering with free and cut-throat competition was a sin against evolution.

[1] For its later influence cf. below, pp. 97–100.

Of the general effect of the *Origin of Species* upon the intellectual and religious life there can be little doubt. If it was as much symbolic of an existing but not yet fully recognised change as instrumental in bringing about that change, in either case its influence was profound. 'To the end of time if the question be asked "Who taught people to believe in evolution?" there can only be one answer—that it was Mr Darwin': that is the testimony of his life-long critic, Samuel Butler;[1] and the concept of evolution thus established has altered the character of man's thought in every department. The immediate shock may have been exaggerated; its lasting consequences have been enormous and universal.

On religion more than elsewhere its effects have probably been exaggerated. Huxley was always an anti-clerical, in spite of his respect for Kingsley and many other clergy. His dislike was partly social—a resentment of the status of scientists: 'Men like Lyell and Murchison were not considered fit to lick the dust off the boots of a curate: I should like to get my heel into their mouths and scr-r-unch it round'[2]—an outburst which, in alluding to the courtly Lyell and the arrogant Murchison, is almost comic. It was also mixed up with his thoroughly Victorian attitude towards women—an inferior and contemptible sex, whose domestic virtues and tenderness for bruised warriors lifted them from degradation, but who were incurably 'parson-ridden' and superstitious.[3] But although he was fond of declaiming against clerical intolerance and rightly set himself to win freedom for scientific research, there is plenty of evidence in his own writings that the hostility of Churchmen was nothing like so violent or general as he had expected. 'I confess I have been pleasantly disappointed,' he wrote in 1863 to the Rev. C. H. Middleton; 'there has been far less virulence and much more just appreciation of the weight of scientific evidence than I expected—and that satisfactory state of

[1] *Life and Habit*, p. 277.
[2] 'Victorian Memoirs' by Lord Ernle, *Quarterly Review*, April 1923, quoted by J. Y. Simpson, *Landmarks in the Struggle between Science and Religion*, p. 193.   [3] Cf. *Life of T. H. H.* I, pp. 211, 262, 387, 417.

things is due, I doubt not, to the much wider dispersion than I imagined of such liberal thought as is manifest in your letter.'[1]

Nevertheless, though the cleavage between religion and science is by no means solely due to the *Origin of Species*, the book's influence has been so great as to require closer examination.

Huxley was undoubtedly right when he fixed on its effect upon the argument from design as the most widely recognised and resented. Ever since Ray's *Wisdom of God* it had been a chief task of the scientist to collect examples of the adaptation of form to function, of the achievements of instinctive behaviour and of the interdependence of living organisms upon one another as proofs of the care and skill of the Creator. Paley had elaborated a similar series of such observations in his *Natural Theology*, and had developed from them a strong teleological argument for the existence of God. This had, in Britain at least, become by far the most popular basis of theism: to attack it was to violate the holy of holies; Darwin, both by his insistence upon 'chance' variation and upon the 'natural' selection of those that survived, seemed to strike both at the element of purpose and the element of deity.

It must be confessed that Darwin's own mind was completely befogged on the subject. He discussed it in many of his letters; for both Lyell and Gray were convinced of the evidence for design. 'I grieve to say', he wrote to Gray, 'I cannot honestly go as far as you do about Design. I am conscious that I am in an utterly hopeless muddle. I cannot think that the world as we see it is the result of chance, and yet I cannot look at each separate thing as the result of Design.'[2] Or again of Lyell: 'I have asked him whether he believes that the shape of my nose was designed. If he does, I have nothing more to say. If not, seeing what fanciers have done by selecting individual differences in the nasal bones of pigeons, I must think it is illogical to suppose that the variations which natural selection preserves for the good of any being, have

[1] From an unpublished letter dated 2 June 1863, preserved in the library of Christ's College, Cambridge, and referring to his *Man's Place in the Universe*. [2] *Life of C. D.* II, pp. 353-4.

been designed. Yet I know that I am in the same sort of muddle as all the world seems to be in with respect to free will.'[1] He certainly was:[2] but it was the honest bewilderment of a man quite untrained in any sort of philosophy or abstract thinking.

Huxley, in the chapter on the reception of the *Origin of Species* contributed to the *Life of Darwin,* is on the contrary clear and vigorous. The idea of chance, if this implies something outside the sequence of causation, he naturally ridicules: chance variation can mean no more than variation whose cause is unknown. Of teleology he sharply distinguishes the 'commoner and coarser forms' and 'the wider which is not touched by the doctrine of Evolution but is actually based upon the fundamental proposition of Evolution': of the former he says, 'The teleology which supposes that the eye...was made with the precise structure it exhibits, for the purpose of enabling the animal to see, has undoubtedly received its death-blow'; and of the latter, 'It is no less certain that the existing world lay potentially in the cosmic vapour and that a sufficient intelligence could, from a knowledge of the properties of the molecules of that vapour, have predicted, say, the state of the fauna of Britain in 1869'. He goes on, 'The teleological and the mechanical views of nature are not, necessarily, mutually exclusive. On the contrary, the more purely a mechanist the speculator is,...the more completely is he thereby at the mercy of the teleologist who can always defy him to disprove that (the machine) was not intended to evolve the phenomena of the universe.'[3] He ends up by insisting that Paley,[4] the 'acute champion of teleology', admitted that actual production might be the result of causes fixed beforehand.[5]

[1] *Life of C.D.* II, p. 378.
[2] Thus, *ibid.* III, p. 189, he welcomed eagerly Gray's assertion of 'Darwin's great service to Natural Science in bringing back to it Teleology'.
[3] *Life of C. D.* II, p. 201.
[4] He quotes *Natural Theology,* ch. XXIII.
[5] For an excellent and recent discussion of this subject cf. Tennant, *Philosophical Theology,* II, pp. 78–120.

A STORM IN A VICTORIAN TEA-CUP

It is obvious that this attitude is vastly more reasonable and, from the standpoint of religion, more conciliatory than the wholesale repudiation of the idea of purpose and of the legitimacy of any form of teleology which characterised the utterances of too many later scientists. But for the plain man it was precisely the appearance of design in structures like the eye, or processes like flight, or chains of behaviour like the parasitism of the cuckoo[1] that had carried conviction. Here were achievements dependent for their success and therefore for their survival value upon the simultaneous and correlated modification of very many factors. It seemed impossible that they could have been built up step by step by a long succession of gradual modifications, particularly as in many cases such single modifications would be not merely useless but harmful. Indeed, in the more complex of them, the story of the young cuckoo for example, the interdependence of the several, and totally different, acts of the drama is so close and each of them so necessary to the result, that Darwin's explanation is literally as absurd as the supposition that a fortuitous coincidence of letters was responsible for the appearance of *Hamlet*. That Darwin himself never faced this issue;[2] that he naïvely assumed that the production of a pouter-pigeon was a complete analogy to it;[3] and that when pressed he merely admitted bewilderment without for a moment recognising the force of the objection; these are facts fully in harmony with the childlike simplicity of his

[1] Cf. below, p. 93, n. 5. The most recent treatment, E. C. Stuart Baker, *Cuckoo Problems*, contains interesting, but in view of p. 56 by no means conclusive, evidence that egg-coloration in the parasitic cuckoos is being affected by natural selection. It does not show that the interlocked sequence of events which together make up the parasitism of *C. canorus* has been or could be achieved on Darwinian lines: that is, it does not deal with the problem of simultaneity.

[2] Cf. e.g. his letter to Fabre, *Life*, III, p. 221.

[3] Cf. e.g. *Ibid.* p. 136 (of greyhound) and the references on *ibid.* pp. 244–5, where he actually asserts an analogy between a Gaucho's skill with the knife and a sandwasp's paralysing of a caterpillar, saying of the Gaucho, 'The art was first discovered by chance, and each young Gaucho sees it and learns it'.

character. But that his successors shut their eyes to the evidence, ignoring what was inconvenient or dismissing it on quite flimsy pretexts,[1] is largely due to the fact that biology, then beginning to confine itself to study of the dead creature in a laboratory, was losing contact with field-work and the observation of the living organism.[2]

It is to the widespread though ill-founded belief that 'Darwin destroyed teleology' that the present tragic conviction of the meaninglessness of life is principally due; and this, though by no means the whole consequence of the conflict, is the first and most disastrous. 'Behold, how great a matter a little fire kindleth!'

[1] E.g. the brushing aside of the conclusions of Fabre (which are fatal to Darwinism) on the trivial ground that the Peckhams, repeating some of his observations, have shown that the accuracy of his insects is not so inerrant as he declared!

[2] Cf. Dampier, *l.c.* p. 301, on the defects of 'laboratory morphologists'.

IV

AFTERMATH: THE RAVAGES OF WAR

> Formerly I was led to the firm conviction of the existence of God and of the immortality of the soul....But now the grandest scenes would not cause any such convictions and feelings to rise in my mind. It may be truly said that I am like a man who has become colour-blind.
>
> CHARLES DARWIN, *Life and Letters*, I, pp. 311–12

1. THE CONFLICT AND THE TRUCE

WARFARE is always disastrous as a method of solving problems; for both parties to it emerge with their ideas narrowed and distorted and their characters inevitably warped. The conflict over Darwinism, coming at an acutely difficult phase in the transition from the old *Weltanschauung*, had unhappy effects both upon science and upon religion. These must first be summarised.

In the first place it produced an estrangement which, as we have seen, had not and need not now have existed. Scientists, as soon as the value of Darwinism became plain, conveniently forgot Owen and their own obscurantists and, fastening upon Wilberforce as Darwin's sole opponent, made religion responsible for him. It gave them some justification that the *Descent of Man*, and the consequent 'ape or angel' controversy, shocked the orthodox much more than its predecessor. The Biblical and Miltonic view of creation was still almost universal; and though the change from an immediate to an evolutionary creation would have been easy, it was hard to surrender belief in the uniqueness and special origin of humanity—especially as the Darwinian hypothesis allowed for no 'jumps'. The traditionalists had a stronger case, emotionally though not of course logically, against the innovation when its application to man was defined for them.

Secondly, the followers of Darwin proceeded to narrow his

conclusions, to reinforce his theory by supplementary arguments, and to insist that thus it provided a detailed and an exclusively complete interpretation. Haeckel's theory of recapitulation (that in its embryonic stages every organism climbs its family tree in order to be born) elaborating Darwin's references to vestigial structures seemed to give an easy illustration of man's pre-human development. The theory of mimicry, of protective and warning colours, associated with Bates' work on the fauna of the Amazon, supplied a number of still more fascinating examples of adaptation to environment. Weismann, by his proof of the isolation of the germ-plasm, claimed to dispose of all possibility of the influence of Lamarckian factors, use and disuse, such as Darwin had himself not only allowed but increasingly favoured. At the same time the emphasis upon 'Nature red in tooth and claw' which appealed strongly to the sentimentality of the age, was erected by Huxley into a dogma in his *Struggle for Existence in Human Society*,[1] in spite of protests from the Russian zoologist Kessler, and the evidence of other factors in Romanes' *Animal Intelligence*. Huxley's pugnacity and ignorance of field-work explain if they do not excuse his distortion of facts: but the result was to give a picture of the evolutionary process hard to reconcile with Christian ideas of God.

Finally Christians, already disturbed by the controversy, found their distress increased by the parallel development of scientific method in the fields of Biblical and historical criticism. In Britain they were as yet unfamiliar with the critical treatment of Scripture: Baur and Strauss, the German pioneers, were regarded, largely through ignorant prejudice, as sceptics if not atheists; and their works were taboo in orthodox circles. So when in 1860 *Essays and Reviews*, a volume of essays by seven liberal Christians, and in 1862 the first of his books on the Pentateuch by Bishop Colenso were published, the shock was almost shattering. Neither the essayists nor the bishop were in fact saying anything that should properly have been regarded as revolutionary. But, as Browning's allusion

---

[1] 'The animal world is on about the same level as a gladiator's show', *Essays*, IX, p. 200.

in *Gold Hair* shows,[1] 'the candid', and Bishop Wilberforce,[2] regarded such views as subversive. A panic was caused: the evolutionists and the higher critics were lumped together in a common anathema; and for a time at least the churches were committed to a strict traditionalism.

The result was therefore the sharpening and continuance of a strife which affected very wide areas of life and threatened fateful consequences. Education was still almost entirely in the hands of Christians: schools and colleges were religious foundations and their staffs were largely in Holy Orders. But science was becoming far too important to the national welfare to be ignored. If the seminaries of the upper classes could still maintain the classical curriculum undefiled—for what need had a gentleman to bother about inventions or medicine?—the Universities could not permanently resist the demand for chemists and physicists in industry or for the application of research to the work of the farm or the hospital. Moreover, there were men now as in the seventeenth century who were finding a thrill in the exploration of nature which neither religion nor politics could equal; the immense expansion of trade and colonisation, of discovery and invention, gave them unique opportunities and some financial support.[3] Science must be free to pursue its objectives without hindrance; the men who had fought interference in the economic and social spheres would fight it not less vigorously here; and if the opposition came from the parsons—the world had had experience of eccle-

[1]   The candid incline to surmise of late
      That the Christian faith proves false, I find:
    For our Essays-and-Reviews debate
      Begins to tell on the public mind
      And Colenso's words have weight.

[2] For a frank estimate of his life and work cf. J. C. Hardwick, *Lawn Sleeves*.

[3] J. G. Crowther's attempt to relate scientific advance solely to economic conditions derives from Marxist theory rather than from historical knowledge: it wholly fails to account for the careers of men like A. C. Haddon whose struggle against poverty is vividly told by Mrs A. H. Quiggin.

siastical control in the past and had no mind to tolerate it in a new form.

Nevertheless a conflict had better be avoided if possible. The number of scientists who, like Huxley, were spoiling for a fight was comparatively small: many of the most distinguished of them, like Clerk Maxwell, were convincedly Christian, and many more genuinely believed that even if some tenets of the faith were hard to reconcile with recent scientific results yet the time for final decisions was not yet. The definite obscurantists among the religious were much less numerous than is suggested by signatures easily given in moments of panic. Many of the best beloved and most influential clergy were liberals and scientists, Arthur Stanley and the pupils of Thomas Arnold, Charles Kingsley, novelist, naturalist and sportsman, Benjamin Jowett, already almost a legend and a contributor to *Essays and Reviews*, and Hort, the colleague of Westcott, the greatest scholar of them all, who 'examined in the Natural Sciences Tripos in the years in which he delivered his Hulsean lectures on *The Way, the Truth and the Life*'.[1] Many, indeed a majority, of the others were pre-occupied with the rights or wrongs of the ritualist controversies which seemed to them of much greater importance than science.

So controversy slackened and an informal but eagerly recognised truce was proclaimed. Science was concerned, so its champions claimed, only with what could be weighed or measured, with the material or physical aspect of existence. Religion, as its own formularies declared, dealt not with the temporal but with the eternal, with the soul not the body, with values and faith. Ever since Augustine's time the principle of the two contrasted citizenships had been a platitude of orthodox Christianity. Science would gladly confine itself to the secular world or terrestrial state if religion would similarly confine itself to the sacred and celestial. On this showing there was, as Ray Lankester[2]

[1] S. C. Carpenter, *Church and People, 1789–1889*, p. 514.
[2] 'There is no relation in the sense of a connection or influence between Science and Religion.' *Kingdom of Man*, p. 63.

flatly stated, nothing in common between the two; and scientists could get on with their work without interference and Christians could cease to worry about matters irrelevant to their beliefs.

It is difficult for us to realise that such a compromise could be honestly accepted by either party—or at least by those who professed an incarnational theology and a sacramental religion. But in other spheres it is evident that the Victorian age had a singular ability to ignore what disturbed its decorum; it was not consciously hypocritical in matters of sex, and had already established a strict reticence in regard to religion; to be a scientist on weekdays and a communicant on Sundays, even if science and religion contradicted one another, was not necessarily to be either a fool or a knave. There are in fact to-day people who accept Darwinism in the laboratory and Genesis in the pulpit without difficulty; fifty years ago such an attitude was more excusable.

So in England[1] for a quarter of a century a sort of gentleman's agreement was observed. Scientists did not trail their coats at the religious; and Christians did not interfere with or dictate to scientific research. An immense amount of work was done in both fields; for in their special department, the very necessary textual, linguistic and historical study of the documentary evidence for Christianity, Biblical scholars did great, indeed invaluable, research, applying a strictly scientific technique to the solution of problems which bear directly upon the validity of the Christian claim. But in the larger task for which scientists and theologians should have been collaborating, the replacement of the medieval *Weltanschauung* by an outlook worthy of the richest experience of modern man, almost nothing was attempted except by individuals; and even then with little success. Neither in philosophy nor in theology (if the two can be distinguished) was there sufficient

---

[1] In Scotland, thanks to a sounder educational tradition and to the influence of men like Henry Drummond, the position was much less obscurantist.

work of a quality in keeping with the excellence of contemporary research.[1]

2. THE CONSEQUENCES FOR SCIENCE

This failure and the isolation of science from religion which was at once its cause and its result had (in the long run if not immediately) serious consequences for both studies. The conflict over evolution had inevitably hardened what had been a theory into a creed; *Darwinismus* speedily became sacrosanct, a cult, almost a religion; and its supporters began to assume the pretensions, and the limitations, of a hierarchy. Discussion was replaced-by eulogy; criticism came to be regarded as disloyal; and conclusions originally put forward with hesitation and subsequently challenged by competent authorities were asserted with a scarcely creditable evasion of their difficulties. The matter deserves examination.

In the first debates on the *Origin of Species* at least four strong arguments had been brought against its theory. First, W. H. Harvey[2] had criticised the basic insistence upon small and inheritable variations, and had advocated large jumps; Huxley had agreed; but the subject was dropped and Darwin's thesis accepted as axiomatic until De Vries. Secondly, as to the claim that cumulative small variations could establish specific distinctness as proved by interspecific sterility, Owen had argued, rightly, that the analogy from domestic races broke down at this point and Huxley had admitted that no evidence of selection producing physiological species had been given; it has never been conclusively produced. Thirdly, the problem of simultaneity, that for example the eye depends for its efficacy upon the simultaneous operation of very many changes each valueless by itself, was Asa Gray's objection;[3] and biologists have not yet fully faced it. Fourthly, the similar objection derived from instinctive behaviour where

[1] Lest this appear too sweeping, tribute should be paid to the memory of James Ward and of Henry Melville Gwatkin, both of whom combined scientific equipment with theological research, and were eminent as teachers.   [2] *Life of C. D.* II, pp. 274–5.   [3] *Ibid.* p. 272.

there is no sign of step-by-step development—where indeed such development is meaningless—had been raised by J. A. Lowell[1] in the early days and J. H. Fabre repudiated Darwin's theory because of it. To these a fifth may perhaps be added, the criticism of the too easy assumption that adaptation to environment is due to its survival value. This is the familiar claim of Kipling's *Just So Stories*—the tawny lion in the desert, the striped tiger in the jungle, and so on. No one, in view of work like that recorded by H. B. Cott,[2] will doubt the facts of animal camouflage or the value of particular adaptation. But in the fierce debate that has raged over the matter any field naturalist of experience will recognise that there has been much exaggeration and mere theorising on both sides.[3] That such objections—some of them plainly fatal to Darwin's contentions—were ignored was largely due to the exigencies of controversy: that the truth of them was denied is an indication not only that scientists share our human frailty but that biological studies were being conducted on too narrow a method and without sufficient observation of the living organism in its natural environment.

This was indeed the most far-reaching effect of the controversy. In physics and chemistry the laboratory had been a sufficient sphere of research. Results impressive in their scope and importance had been constantly achieved, and the daily life of civilised man testified to their value. It was right and natural that similar methods should be applied to organic structures and biological problems. Biophysics and biochemistry have a great part to play in providing indispensable evidence as to the mechanisms of the living organism. In the problem of adaptation and colour pattern, for example, it has long been evident that the easy theorisings as to mimicry or warning-colours need to be drastically reviewed and constantly disciplined by reference to research into the biochemistry and physiological character of

---

[1] *Ibid.* p. 239.   [2] Cf. his *Adaptive Coloration in Animals*, 1940.
[3] E.g. in the case which J. S. Huxley, *Evolution: the Modern Synthesis*, p. 470, regards as 'the best of all', industrial melanism in Lepidoptera.

pigmentation and the physical effect of environment. But such studies only provide one part of the necessary evidence; and to suppose that life can be wholly explained by them is as absurd as the attempt to interpret Captain Oates' sacrifice at the South Pole in terms of the state of his glands.[1]

Unfortunately, the belief in their omnicompetence was encouraged not only by the opposition of religion but by the circumstances and outlook of the time. 'When the human mind invents or encounters the mechanistic theory of the organism, it is confronted with an apparition which it at once recognises as the darling of its adolescence and the symbol of its power— a machine.'[2] So wrote one of the more brilliant young biologists whose early death prevented him from following up his warning. In a mechanical age economic man and machine-conditioned thought are characteristic products; and the sciences which were responsible alike for commercial supremacy and for an industrialised social order were liable to explain all human activities in terms of materialistic categories. Dr Joseph Needham's book, *The Great Amphibium*, and his subsequent writings show the extent to which even in a Christian's thinking science has become identified with mechanism; and how easily the scientist finds himself living in several different worlds which have nothing in common except that he belongs to them all. Specialisation on the one hand and the dread of 'metaphysics' on the other made the abandonment of the attempt to hold an integrated or synthetic view of experience fashionable. The consequent outbreak of scepticism, of moral anarchy and intellectual irrationalism, is the natural result of such dissociative philosophies.

The prevalence of this self-imposed limitation upon the scope of science is most plainly seen and has had its worst effects in the field of psychology. Ever since it became impossible for scientific

[1] Cf. H. H. Farmer, *Experience of God*, p. 121.
[2] A. D. Darbishire, *An Introduction to Biology*, p. 85: he was one of the first to recognise the irony of the fact that scientists, condemning the anthropomorphism of Christians, themselves fall into mechanomorphism.

materialists to deny the effect of mental impressions[1] upon the physical organism, the controversy as to the status of psychological studies has been continuous. Was psychology to be called a science? To say 'No' would be to allow that the sciences had to take cognisance of influences which upset their results but which they were not competent to deal with. To say 'Yes' would be to abandon the criteria of weight and measurement and to open a backdoor to influences against which the front had been strictly and publicly barred. But if so, what sort of psychology was eligible? Much of it was no doubt scientifically reputable—intelligence tests, examinations of nervous reactions, observations of physical behaviour. But much of it—and all the more popular and exciting departments—were not only speculative and controversial but almost as much beset with charlatans as spiritualism. Where was the line to be drawn? In the early days of the Nancy School, indeed among the orthodox until very much later, psychology was, not altogether undeservedly, frowned upon; and when as a result of its practical success in dealing with the various traumas classified popularly as 'shell-shock' it could no longer be dismissed, its blatant exploitation, and the sensational claims put forward by interested parties, revived earlier suspicions. What was to be done with a subject evidently necessary to the medical man and in some degree at least to the biologist which yet refused to conform to the categories recognised as scientific? The question has hardly yet been finally answered, though the work of Max Wertheimer and of the Gestalt school has gone far towards a satisfactory solution of it.[2]

It is perhaps unnecessary to draw attention to a proposed solution of transatlantic origin, the attempt of Dr J. B. Watson and his school to reduce all human life to the level of Capek's robots. Behaviourism, defining existence as that which could be perceived by the senses and adopting as its basis Pavlov's researches

[1] 'Exteroceptive stimuli' is their scientific title: cf. below, pp. 71–2.
[2] For a clear and brief discussion of this solution cf. K. Koffka, *Principles of Gestalt Psychology*, pp. 13–23.

into the conditioned reflex, identified thought with laryngeal movements; as it admitted that these movements were imperceptible and therefore on its own definition non-existent, the hypothesis was more ingenious than convincing: as Professor Macmurray demonstrates, 'if true, it cannot be true'.[1] The school has made some useful observations and added something to our small knowledge of animal behaviour; but its doctrine may be regarded as the *reductio ad absurdum* of the attempt to claim for science both omnicompetence and a solely mechanistic technique. Obviously for the full study of the living organism such limitation could not apply: biology could not be reduced to biochemistry and biophysics.[2]

3. THE CONSEQUENCES FOR RELIGION

If the effects of the controversy upon science were distorting, they were not less deplorable upon theology. By the theologians of the Darwinian age, notably by F. D. Maurice in the field of philosophy and sociology, and later by Lightfoot, Westcott and Hort in that of textual and historical criticism, a real development of scientific study had taken place. Maurice had vindicated the relevance and creativity of Christianity in the ordering of social relationships, laying the foundations of a synthetic interpretation of Christian doctrine and indicating its consequences for the reform of our industrial and economic system. The Cambridge trio had seen the necessity to establish the true text and authenticity of the evidence for the life of Christ and had made an incomplete attempt to interpret the meaning of the New Testament in the light of the best scholarship of the day.

Unfortunately, few of their contemporaries were interested and

[1] *The Boundaries of Science*, p. 135.
[2] Attempts so to reduce it are still so common that it is worth referring to the striking section in J. H. Woodger, *Biological Principles*, pp. 311–17, in which he demonstrates by empirical data that 'the assertion of those who say that biology must *only* use the concepts of physics and chemistry is not only not practised but not practicable'.

many were hostile;[1] and few of their successors were capable of carrying on the task, except by rather slavish imitations. No really bold attempt was made to complete the study of the records and to use them as the basis for a reformation of Christian doctrine, cultus and organisation. Some of the more foreseeing scholars realised that the new knowledge could not be much longer confined within credal forms of the fourth or a theological schema of the thirteenth (or sixteenth) centuries. A few made attempts to express their convictions in a form appropriate to an age of science, industrialism and world unification. But any radical experiment was condemned by authority and traduced by public clamour; and the majority of Christians preferred 'safe paths in perilous times'.

Moreover, it must not be forgotten that until a generation ago the number of schools in which there was any teaching of science was negligible; and even then the subject was often taken only by boys who were adjudged incapable of success in classics or mathematics. When G. W. S. Howson went to Holt and made it one of the first public schools with a modern curriculum, he left Uppingham where he had been Science Master with a makeshift laboratory in which there was no balance that weighed accurately nor any adequate apparatus. In most of the more famous schools there was nothing except the barest elements of chemistry and physics until well into the present century. In consequence there is even now hardly a single bishop or prominent Christian leader who has ever studied any physical science, and indeed very few who have ever worked in a laboratory. For men unacquainted with the principles and results of scientific research it was perhaps easy to believe that a subject of which they knew nothing could not seriously affect one in which they were professionally interested. Even to-day it is noteworthy that popular theologians from Dean

---

[1] Cf. F. von Hügel, *Selected Letters*, p. 254: 'Dr Pusey was incapable, had made himself incapable, or deliberately acted as though he were incapable, of taking any interest in anything that was not directly technically religious or that was not explicitly connected with religion.' He adds: 'This was quite uncatholic, quite unlike the Jesus of the Synoptists.'

AFTERMATH: THE RAVAGES OF WAR

Inge[1] to Miss Dorothy Sayers[2] make blunders in matters of science which they would not venture to leave unchecked if they were dealing with literature or history.

By the end of the century theology was represented by the Oxford 'greats' tradition,[3] Plato and Aristotle, christianised and liberalised but almost wholly untouched by the influence of the sciences. In scholars like R. C. Moberly, J. R. Illingworth, Charles Gore, H. Scott Holland and E. S. Talbot—the men of *Lux Mundi*—in William Bright, and in the more liberal W. Sanday, there was a general similarity of outlook and a massive learning: of their achievement Dr A. C. Headlam's *Christian Theology*[4] is the survivor and the monument. This tradition represents the doctrine of the Anglican Church, of the 'judicious Hooker' and the more orthodox of the Caroline divines, that is of the patristic age as expressed in Athanasius rather than in Origen or in Augustine, in Cyprian rather than in Stephen or Novatian. It is a *via media* but with a character of its own, orthodox but not intolerant, comprehensive but hardly liberal, systematic but in an easy-going English fashion. It maintains a strong sense of worship and of mystery expressed in the dignity of its liturgy and in its emphasis upon the cultus; and combines this with a practical concern for the cure of souls, the redress of social evils and the extension of missionary effort. At its best it is a worthy representative of Victorian England. But it is almost entirely unaware of the vast changes that have come over human thought and life since the seventeenth century; and is hence unable to integrate and hold together the rapidly disrupting departments of man's vastly complex activities. It is a monument to the past, not a rallying-point for the present, or a beacon for the future.

[1] Cf. e.g. his remarks on the 'intellect' of insects in *The Fall of the Idols*, pp. 49, 50: these are perhaps derived from a casual reading of R. W. J. Hingston, *Problems of Instinct and Intelligence*.

[2] E.g. on 'the Mendelian law'. *The Mind of the Maker*, p. 4.

[3] For an interesting and independent expansion of this cf. Adolf Löwe, *The Universities in Transformation*, pp. 5–10.

[4] Published in 1934 but evidently written some twenty years earlier.

AFTERMATH: THE RAVAGES OF WAR

Disintegration and specialising are thus the most disastrous result of the conflict between science and religion. At a time when the expansion of human interests was so swift and when the field of knowledge was becoming too large for any one mind to master, only the full co-operation of the representatives of science and religion could have preserved a sense of the unity of all truth and of proportion in its service. Departmentalism was inevitable; for it took a lifetime of study to attain eminence in any line of research:[1] but if the leaders in the two great interests had not been estranged and driven into an angry exaggeration of their differences, the dangers of extreme specialisation might have been avoided. As it was theology perforce abandoned its ancient status as 'queen of the sciences'; slipped contentedly into studies appropriate only to archaeology or history; and came to devote more and more research to matters of less and less intrinsic value. Any sense of the relationship of these studies to the whole body of human knowledge began to disappear; learning became the hobby of the expert; a jargon of esoteric terms came into use; and as at the Renaissance so now the subject gained a reputation for irrelevance, meticulousness and pedantry.

For a decade and more the danger of this state of locomotor ataxia in which humanity ceased to obey any coherent and organising control was concealed by a blind belief in progress—a strange irrational faith in the automatic improvement of human life, based only upon a shallow view of evolution and a blinkered ignorance of what other men in other fields of effort were doing. The experts in each department were so learned, the development of discovery and invention was so sensational, the enrichment of civilisation was sò impressive, that life must surely become safer, healthier, happier, more divine. The instruments for such improvement were available: there was neither the knowledge, nor the trust, nor the energy to use them. The period ended in the bloodbath of the first Great War.

[1] The tradition of omniscience died out with Mahaffy in Dublin and Gwatkin in Cambridge—though at St Andrews Sir D'Arcy Thompson may perhaps still maintain it.

V

THE 'NEW REFORMATION': CAN WE ACHIEVE IT?

The misfortune that has overtaken the spiritual outlook of man is that as his universe expanded his conception of the deity did not expand with it.

F. WOOD JONES, *Design and Purpose*, p. 75

1. SIGNS OF HOPE

THE 'new Reformation' which, as Huxley claimed, must necessarily follow the scientific Renaissance has not yet come. Mankind, forced into one neighbourhood by the annihilation of geographical barriers, has not found the common outlook and ideals which are essential to neighbourliness. In consequence since a stranger on one's doorstep is more annoying than a stranger miles away, the good gifts of applied science have made not for peace but for war. Nor is there any prospect of an end to strife unless we achieve either a common loyalty which shall weld us voluntarily into unity or a common enslavement under a tyranny that we cannot resist; and the latter will of course only postpone or change the character of the battle. An indispensable preliminary to the acceptance of a common loyalty is the reconciliation of science and religion which a new Reformation would demand and enable. We have seen that at the end of last century the prospects were bleak. What are they now?

Before attempting an answer two new factors, in physics and in psychology, must be noted.

A. THE NEW PHYSICS: EINSTEIN, RUTHERFORD, HEISENBERG

Until recently by all of us and even now by very many experience is sharply divided into objective and subjective, the facts that we can test in a laboratory and the ideas and values which belong to our mental and personal quality. In the great tradition of

64

THE 'NEW REFORMATION'

European philosophy derived from Plato it is an axiom that the world of thought is the real world and the world of sense-perception is derivative from it, a reflection more or less distorted, an 'outward and visible sign of an inward and spiritual grace' as Christianity expresses it. Since Descartes, and increasingly with the stress laid by scientists upon quantitative evidence, a reversal of this axiom has become popular—not so much with the more philosophical but with the unimaginative and the technicians. Thought itself came to be treated as a sort of by-product or epiphenomenon; and material objects were regarded as facts, while values were dismissed as fancies. It seemed clear that man lived in and could largely control a material world whose structure and functioning he could measure with absolute accuracy, a world which was what his perceptions aided by scientific instruments demonstrated it to be.

Nobody (or at least nobody in his senses) ever believed that his mother's love for him or his own affection for his friends was less 'real', less actual and significant and effective, than the chair on which he was sitting or the test-tube with which he worked. But then nobody, whatever his enthusiasm for materialism, has ever genuinely regarded himself as a robot. Yet in the days before Einstein there were plenty of us who were prepared to argue not only that all values were subjective and (in the old sense of the word) accidental but that whereas they varied with the tastes of the individual the material world existed in the form in which we knew it and that this form was wholly objective and independent of humanity. To argue that it was a closed system, mechanical in its construction and inexorable in its routine, was a usual consequence.

Is it still necessary to insist that this naïve belief has fallen hopelessly to pieces? Even those who like myself are quite incapable of appreciating the proofs of relativity must be aware that such confidence in the objectivity of the earth and the accuracy of man's measuring-rods is no longer warrantable. If not, let them re-read Sir Arthur Eddington's Gifford Lectures with the account of his two tables and of the contrast between them or of his entry into a room.[1] Whether or no they agree with his philosophy,[2] at least this picture of matter is radically different from that of the nineteenth century. The atom is no longer indivisible and indestructible; measurement is relative to the measurer; as in the eighteenth century so now there are two alternative and seemingly contrasted theories as to the character of light; far from the system and course of events being determined 'the denial of determinism is not merely qualitative it is also quantitative; we have actually a mathematical formula indicating just how far the course of events differs from complete predictability'.[3] Man instead of being the measure of all things and the lord of creation discovers himself to be capable only of forming a subjective, fragmentary and, for all he knows, fantastically inexact concept of a universe immeasurably transcending his powers of appreciation or of control. What scent particles are decoded into perceptions by the pectinated antennae of the male oak eggar? How does the eel steer its course through unknown seas from an English river to its spawning-ground in the West Indies? What would the world look like if our eyes could see infra-red or ultra-violet as colour? What would be our sensations if we could 'pick up' directly the wave-length of the 'Forces Programme'? Consider the range of vibrations from which our senses construct the world as we perceive it and contrast this with the ranges to which we are insensitive, and the need for humility if not for agnosticism will become plain. It *is* a mysterious universe, and dogmatism is indecent.

That does not involve a thorough-going scepticism. Because we are human and our eyes are not 'a pair o' patent double

[1] *The Nature of the Physical World*, pp. xi–xiv and 342.
[2] Before doing so it is well to read Professor Susan Stebbing, *Philosophy and the Physicists*—though to avoid exaggeration her preface should be given full weight.
[3] Eddington, in *Science and Religion* (1931), p. 125; cf. his *Philosophy of Physical Science*, pp. 90–1.

million magnifyin' gas microscopes of hextra power' that is no
reason for shutting them or behaving like beasts. We are men with
the limitations and capacities of our species; and if our science is
relative, not absolute, we nevertheless have to live by it. Indeed,
what has been destroyed is not the knowledge that we possess—
this remains valid and valuable—but the belief in its objectivity
and finality; and this, though philosophically of enormous significance, makes little practical difference. If we now recognise that
there is an element of indeterminism and unpredictability in our
calculations, this does not affect their general reliability. The
measuring-rod of a man is sufficiently accurate for the human
craftsman and engineer, even if its records are no longer ultimately
exact. What has happened is that probability here as in other
fields takes the place of certainty; laws become hypotheses and
infallibilities disappear for science as they ought to have done for
theology.

But this of course means that the mind of the measurer has
become once again more important than his instrument, which
is inexact, or the object measured, of which he gets only a subjective impression. If we are not driven to a Berkeleian idealism
in which mind alone has any real existence or to Eddington's own
'selective subjectivism', this can only be because, as Professor
Stebbing puts it, 'the physicist's assertions are always to be treated
by reference to sensible experience: he begins from and returns
to sensible experience';[1] he cannot by changing his picture of it
escape from the world which common sense and daily verification
compel us to accept even if we are less cocksure than our forebears
as to its structure and character. That our own mental processes
create what our senses perceive, may be arguable and is perhaps
impossible wholly to refute: but Dr Johnson's crude refutation[2]
is borne out not only by the general orderliness of the universe but

---

[1] *Philosophy and the Physicists*, p. 27; cf. also p. 66.
[2] 'Johnson answered, striking his foot with mighty force upon a large
stone till he rebounded from it, "I refute him thus".' Boswell's *Life*
(edit. 1900), I, pp. 347–8.

by the plain conviction that whatever we may say of tables and trees other human beings whom we know and love or hate have a self-evident objectivity. 'No one believes in solipsism, and very few even assert that they do.'[1]

But if we grant the reality of the world of science on evidence of this kind, we are no longer in a strong position to deny the reality of the world of history or of religion. The perceptive faculties of the poet or the mystic are not exercised upon the same objects or order as those of the physicist or the biologist: they use different instruments and a different technique of interpretation; but to say on *a priori* grounds that the result is less true is as unjustifiable as to say that it is less valuable. This may be the case: science can test and corroborate its results more easily, that is because they stand lower in the scale of being, are less complex and less personal. But if science is committed to quantitative investigation, it is thereby debarred from dealing with the more typically human experiences and must not cavil at other enquirers whose concern is with qualities and personalities. 'Science', said Gilbert Chesterton, 'can analyse a pork-chop and say how much of it is phosphorous' (*sic*, I regret to say) 'and how much is protein; but science cannot analyse any man's wish for a pork-chop, and say how much of it is hunger, how much custom, how much nervous fancy, and how much a haunting love of the beautiful. The man's desire for the pork-chop remains literally as mystical and ethereal as his desire for heaven.'[2] In 1905 those words were very necessary. They are still timely for such scientists as have not learnt the lesson of relativity and are arrogant enough to claim that their knowledge is of objective fact whereas that of the theologian is of subjective fancy. But for the rest of us, whether scientists or theologians, the sharp distinction is no longer valid:

[1] Eddington, *The Philosophy of Physical Science*, p. 193: this book is the most recent and complete statement of its author's characteristic position —a position which cannot be dismissed so easily as a hasty reading of Professor Stebbing may suggest, or as some far less qualified scientists try to dismiss it.   [2] *Heretics*, p. 146.

it is in both cases with a man's subjective and to that extent mystical relationship to the pork-chop that we are concerned, whether we are discussing its chemistry or its edibility or its aesthetic appeal.

B. THE NEW PSYCHOLOGY: FREUD, ADLER, JUNG

Development in the field of psychology has accompanied and assisted this change. Started by James Ward's article in the ninth edition of the *Encyclopaedia Britannica* in 1886 this phase of the subject had been vigorously expounded by William James, G. F. Stout and W. McDougall; and in Britain at least a sound and scientific approach to psychological studies had become established before the impact of Freud's work and the vast vogue of psychotherapeutics began to create popular excitement. As a cult the 'new psychology' had immense popular assets, esoteric jargon,[1] pornographic appeal, grandiose claims and some measure of achievement. Nevertheless, despite its exploitation by lecturers, practitioners and publicists whose qualifications were often almost non-existent—an exploitation which took its rise from the prevalence of 'shell-shock', neurasthenia and sexual disorders after the Great War—its influence was not so damaging as it might otherwise have been. The subject was discredited indeed, but along with much exaggeration and some charlatanry good work was done and steady progress made. If the Viennese schools have made a greater contribution to medicine and pathology than to epistemology and philosophy, yet methods of more strict research have checked and tested their claims, and won for the subject a generally recognised status as a branch of science.

Of its importance in relation to the general scope and content of scientific study and particularly as providing a line of enquiry into the problems raised by the new physics there can be no

---

[1] An example of the persistence and difficulty of this is seen in the long discussion of the meaning of Super-ego in Waddington, *Science and Ethics*, pp. 56–88, from which it appears that Freud never used the term consistently and that his successors disagree about its meaning.

question. In a universe in which we can no longer find absolute objectivity, the mind of the investigator, the character of the instrument by which he records and measures, become issues of primary significance. Man's relationship to the pork-chop is a matter on which psychology cannot be refused a hearing. The range which our consciousness can cover, the mechanisms by which we register our impressions, the processes by which results are tested, compared and manipulated, the validity which can be assigned to our conclusions, these are now indispensable aspects of scientific enquiry. Indeed, epistemology, in this age of relativity, takes on a fresh and vital status; and for it psychology is a necessary equipment.

That this involves an extension of the customary limits of science, a surrender of the detachment which it has traditionally demanded, is obvious. Not only so, but, as recent developments in psychology demonstrate, the study of the individual and of the origin and character of uniqueness is becoming rightly prominent;[1] and the effect of this is to modify profoundly the supposed scope of scientific research. Indeed, it is largely the acceptance of psychology as a science, complementary as this acceptance is to the stress upon the subjective and relative elements in the new physics, that has led the younger men to claim all organised knowledge as the scientist's proper province, and to consider how methods can be developed for the scientific study of ranges of experience not amenable to the technique appropriate to material objects in a laboratory. That little progress has yet been made is due as we have seen partly to the divorce between field-work and research; and in particular to one result of this, the lack of any adequate knowledge of evolutionary or comparative psychology.

A thorough-going evolutionism, such as we have been advocating, will of course insist upon a continuity between man and the animal creation more complete than Darwin's generation could be expected to accept. To recognise in ourselves the physical

[1] See for example the concluding chapter of G. W. Allport, *Personality; a Psychological Interpretation* (1938).

and psychic evidences of our ancestry is only half the task: we ought also to trace the origin and development of our psychic and mental characteristics by as thorough a study of the evolution of behaviour as has been devoted to the evolution of structure. This has not yet been done. There exist a few valuable monographs mostly devoted to the higher apes; a mass of detailed experiments under artificial conditions such as Thorndike, Pavlov and others have recorded (the value of which is much less than is usually claimed); a few admirable and very many worthless studies and observations of animal and bird habits; and a huge and miscellaneous collection of more or less casual data in which there may well be some useful material. But the subject has not yet received its Darwin; and, as the few books which profess to deal with it suggest, there is probably need of more numerous and more exact field-workers before any sound genetic treatment is possible.

In consequence the study of human conduct lacks secure foundation; and psychology has depended far too largely upon methods of analysis which are often dangerously subjective and in any case mainly devoted to the abnormal and the diseased.[1] A true picture of man's psychic nature cannot be reached unless there is far more accurate knowledge than we at present possess of his inheritance. Presumably he has vestiges at least of those instinctive achievements which the lower levels of life disclose— of the immediate sensitiveness to a common impulse which makes a flock of dunlin behave like a single bird, of the sense of direction which brings the shearwater released at Venice back at speed to its nesting-place in South Wales,[2] of the response to exteroceptive stimuli whereby if the egg laid daily be removed the wryneck, and

---

[1] Thus e.g. G. W. Allport, *l.c.* pp. 215–16, lists and condemns as inadequate 'in dealing with the subtle characteristics of genuinely mature personalities' all the thirteen existing schools or branches of psychology concerned with personality; and recent news suggests that even the Gestalt school, the most impressive of them, is breaking up.

[2] Cf. D. Lack and R. M. Lockley in *British Birds*, xxxi, pp. 242–7.

THE 'NEW REFORMATION':

many other birds, will go on laying vastly beyond the normal limits of the clutch.[1] Until we know the character of such capacities and their place in the development of living organisms, we have not got the material for a fully consistent interpretation of the evolutionary process. But that the opportunity and need for such studies should be widely recognised shows how far we have moved from the days when man was supposed to be linked with the rest of creation physically but to be wholly *sui generis* in his psychic and mental attributes.

2. CONSEQUENT INTEGRATIVE PHILOSOPHIES

In consequence of the breakdown of the materialistic concept of the universe the more human and intelligent scientists (and to a less extent also theologians) have begun to repudiate the truce and its imposed frontiers. They recognise that though it may be convenient to call knowledge when applied to one set of data science and to another scholarship and to another philosophy, this must not imply a radical division or any rigid distinction of scope or technique. We have to live as integrated personalities; for if psychology has taught us (as yet) less than at one time it claimed, at least it has proved that integration is essential to effective and healthy life. We have become aware that modern industrial society produces an exaggerated specialisation of function and tends to reduce mankind to the level of Mr H. G. Wells' selenites. We therefore welcome the trend apparent in all political and corporate effort towards co-operation and unity. And the result is a number of essays and movements aiming at a synthetic interpretation of the universe.

Probably the most important of these in its effect upon the history of thought will prove to be the series of books in which Henri Bergson set out his highly suggestive system of philosophy. His early work, and especially the fascinating *L'Evolution Créatrice* published in 1907, was powerfully influenced by J. H. Fabre's

[1] Cf. F. H. A. Marshall in *Phil. Trans.* B, 1936, vol. 226, p. 443.

studies of insect behaviour—studies which seemed to their author and to Bergson wholly irreconcilable with Darwin's theory of selection operating upon small changes. The interdependence and therefore simultaneity of the developments manifested in any elaborate piece of structure or behaviour—the vertebrate eye, the achievement of flight, the breeding-habits of the predatory wasps, the parasitism of the cuckoo—had been noted by Darwin and Asa Gray but ignored by their successors; and the new study of mutations, though allowing for large jumps in evolution, did not indicate any idea as to the stimuli or the mechanisms by which such complexities could be evolved. Bergson's own myth of an *Elan vital*, reviving the ancient belief in an *Anima mundi* and the careful and interesting theory of a Plastic Nature, expounded by Ralph Cudworth the Cambridge Platonist and John Ray the great naturalist, stresses both the continuity and the creativity of the evolutionary process, and has obvious links with the undeveloped but continuously held Christian doctrine of the Holy Spirit. Bergson's work, though popularised (and much cheapened) in England by Bernard Shaw and H. G. Wells, has not yet received adequate consideration from scientists or theologians, although when completed by his last work, *Les Deux Sources*, it represents the most important and original contribution to philosophy in recent times. It is still too true that a man who can make philosophy attractive is always suspect in academic circles.

Fifteen years later the importance of a synthetic interpretation of the scientific outlook in relation to philosophy and to human welfare was emphasised by the appearance almost simultaneously of five books of real importance. In 1922–3, C. Lloyd Morgan, who had been a pupil of Huxley and a pioneer in the study of animal psychology, delivered his Gifford Lectures which appeared as *Emergent Evolution* (1923) and *Life, Mind and Spirit* (1926). In 1925, A. N. Whitehead, the mathematician and philosopher, delivered and published his Lowell Lectures, *Science and the Modern World*. In 1928, Sir Arthur Eddington published the first of his

73

expositions of the philosophical effects of relativity, *The Nature of the Physical World*. In 1929, Dr J. S. Haldane published his Gifford Lectures, *The Sciences and Philosophy*, in which he followed up his earlier volume, *Mechanism, Life and Personality*. To these may be added the very remarkable book, *Holism and Evolution*, published in 1926 by Field-Marshal J. C. Smuts, whose interest in philosophy and experience as a field-naturalist, added to his clarity of mind and range of practical knowledge, form an excellent qualification for work of this kind.[1]

On the side of theology there was for a short time a real hope of a reformation.[2] The vigorous practical efforts originating with those whose service overseas in the mission field or in the Great War had given them a wider horizon and a measure of freedom from the traditional orthodoxies of Europe had drawn Christians into a study of other religions, of social and economic conditions and of their own faith which provided the occasion for fresh and relevant thinking. One very great book, John Oman's *The Natural and the Supernatural* (1931), and a number of other studies of the same fundamental problem[3] were published; and it seemed possible that the changed attitude of science might be met by a corresponding change and a willingness to face the common task sympathetically. When Oman summarised his conclusions in the words 'Reconciliation to the evanescent is revelation of the eternal and revelation of the eternal a higher reconciliation to the evanescent',[4] he stated the principles of a theology in which there could be no ultimate antithesis between nature and grace or between science and religion, in which indeed the worlds of the scientist and the theologian were seen to be one and the same,

---

[1] I have discussed these at some length in my book, *The Creator Spirit*.

[2] A judicious treatment of what he calls the Wider Immanentism is in E. C. Dewick, *The Indwelling God*.

[3] The most important being Dr F. R. Tennant, *Philosophical Theology*, 2 vols. 1928 and 1930; Dr W. Temple, *Nature, Man and God*, 1934; and B. H. Streeter, *Reality*, 1926.

[4] *The Natural and the Supernatural*, p. 470.

their unity being sacramentally or incarnationally interpreted. On that basis the 'New Reformation' could—and in some respects actually did—proceed.

For Oman was expressing the convictions which underlay the very significant efforts of the Protestant churches of Europe and America to bring religion into more vital relationship with contemporary life and thought; efforts which arose out of the concrete needs of those who were trying to interpret Christianity to the peoples of Africa and Asia; efforts which gave rise to widespread and serious discussion both of the 'Life and Work' and of the 'Faith and Order' of the Church.

Starting from the Edinburgh Conference of 1910 there had been a tardy and timorous but nevertheless honest and penitent recognition of the extent to which the Church had failed to use the best knowledge of the time, to criticise and control social and racial development, to emancipate itself from ancient abuses in thought and practice. Faced with actual problems which demanded fresh ideas and methods for their solution Christians were compelled to investigate their traditions, to modernise their ethics, and even to begin to re-state their theology. The Conferences in Birmingham and Stockholm, in Lausanne and Jerusalem, if they did not go very far in achieving reforms, at least asserted the conviction that large-scale changes in the interpretation and the application of Christianity were inevitable and brought together a body of scholars and students not ill-qualified for the work. Ten years ago it seemed as if the conclusions which Oman had reached would win general acceptance and might well initiate a new era; and to some of us that hope still remains—though we hardly expect to live to see it actualised.

3. OBSTACLES AND REACTION

Unfortunately the times are evil. As the breakdown of civilisation in the fifth century and the hideous warfare of the sixteenth produced Augustine and Calvin, so the calamities of 1914–18

75

produced Dr Karl Barth and the new Calvinism with its insistence upon the disparity of nature and grace, the futility of all human effort, and the total depravity of man's fallen state. As a protest against the shallow and over-optimistic confidence of some American humanists, as a reminder that progress has never been automatic or identical with material success, as a call to a deeper appreciation of evil and of man's need and as an incentive to humility and adoration, Dr Barth's work was valuable. But his own subsequent recantation has proved that those who asserted that it was both exaggerative and immature were not unjust to him.

His school, which in Britain and America was neither impressive in quality nor (at first) strong in numbers, would have mattered little if it had not coincided with and been able to adopt the familiar Hegelian schema of dialectic. This formula, so attractive to minds which like a simplified and regimented existence, to minds like those of Marx or Hitler which see the world in terms of crude contrasts, had already been introduced into popular theology by Albrecht Schweitzer's brilliant book *Von Reimarus zu Wrede* (*The Quest of the Historic Jesus*) in which a century of theological study had been forced into a dialectical shape and the principle of *entweder-oder* been raised to the level of an axiom. To men and women exhausted by the strain of a world war and bewildered by the sheer complexity of modern life, the beautiful notion that life is all white met by and opposed to its antithesis that life is all black seems both restful and satisfactory. It can then be laid down that the only solution is a paradox (which too often means that any paradox is a solution); that tension is creative; and that all thought must be revolutionary, swinging violently from one extreme to another. To see the world as a chessboard with yourself and friends white and everyone else black is no doubt soothing: it gives you just cause to denounce or to liquidate. But even if, in moments of inverted pharisaism, you see yourself black and accept the consequent damnation, the picture does not become more true to life. For with every respect to Augustine, Calvin and Barth life is multi-coloured—and in any case Jesus said 'Judge not'.

This paradox-mongering or theology of crisis represents a phase of development which is perhaps normal for the young. Many of us passed through it when G. K. Chesterton—that grand master of paradoxes—was in his prime. It meets the youngster's desire to say 'my friends are exaltations, agonies', to be rapt into the heavenlies at one moment and plunged into hell the next. Both these experiences are necessary to maturity; and life is the richer by the extension of its range. But to exist in a state of violent transition, to refuse to reach any stability or equilibrium, to deny the final member of the Hegelian triad, is to fail of adjustment to life or to eternity.[1]

Extravagance, not to say exaggerated obscurantism, is unhappily of the essence of this reaction. At a time when, as we have seen, the abler minds in Christendom were beginning to respond to the new outlook in science these 'revolutionaries' reverted to the legends and superstitions of the old order or to an irrationalism which contentedly writes meaningless rhetoric and justifies it as the proper language to apply to the ineffable. The reasonable was renounced in favour of the numinous, the ethical and prophetic in favour of the apocalyptic and eschatological. 'Liberal', 'rational', 'modernist' became terms of abuse. History was made subordinate to facts never actually observable and to theological ideas current in the early Church. Problems of date, authorship and authenticity were brushed aside not merely as insoluble but as irrelevant. There were abundant signs that Christianity was being reduced to the level of an improved Mithraism; or even that the apologetic rejected a generation before—'the Universe is illogical:

[1] Very characteristic is Reinhold Niebuhr's insistence upon anxiety and uneasiness as essential qualities of the Christian; cf. *Nature and Destiny of Man*, I, *passim*. Equally characteristic is his habit of dismissing what he dislikes on the ground of its illogicality and defending his own preferences on precisely the same ground. To be paradoxical is for him a virtue in Kierkegaard but a vice in Spinoza.

THE 'NEW REFORMATION':

Christianity is also illogical: therefore Christianity explains the Universe'[1]—was being put forward as the latest version of Catholic orthodoxy.

With the economic catastrophes and the increasing threats of war during the subsequent decade such reaction distracted the attention of religious thinkers from their proper task and destroyed the hope of a reformation. The outbreak of the struggle in 1939, though it plainly demonstrated the need and the opportunity for the assertion of an integrating and, as many recognised, authentically Christian ideal, found the Church theologically and ethically divided and in consequence practically impotent. And when scientists acknowledged their need for a new *Weltanschauung* which should serve as a basis for practical citizenship and a future programme, their requests, tentative indeed and inarticulate, met with little or no response. It was left to the bright superficiality of the Brains' Trust to supply the public with answers to its problems and guidance in its search for a philosophy of life.

Indeed, since the outbreak of war such theology as has been produced has been plainly pathological, the utterance of men who, having solemnly declared in 1937 at the Oecumenical Conference at Oxford that war was a demonic evil, an evil which no emotional fervour must be allowed to extenuate, found themselves constrained within a couple of years to do evil or at least to advocate the doing of evil that good might come. Their position is not enviable; and it is hardly surprising that a subconscious awareness of its falsity has produced a distorting effect upon their thought. In self-defence they are bound to emphasise the universality and power of evil, until Christianity becomes a message not of salvation but of the Fall;[2] until Satan becomes (as with John Milton) hardly less powerful and much more interesting than God;[3] and

---

[1] Cf. W. H. Mallock, *Reconstruction of Belief*: he was a dilettante who joined the Church of Rome on his death-bed!
[2] Cf. D. R. Davies, *Secular Illusion and Christian Realism*.
[3] Cf. C. S. Lewis, *Broadcast Talks*.

78

until an exaggerated enthusiasm[1] exalts the strange and diseased genius of Kierkegaard[2] into the place of the most profound Christian theologian. The Church's faith in the Holy Spirit, the Pauline emphasis upon good news and charity and fellowship, the message of the Johannine Prologue, indeed all the characteristic passion and power of the religion of the New Testament have disappeared. In their places such writers give us only the doleful news that 'the world is very evil: the Church has always known it'. And this is not the gospel—or even its proper prelude.

[1] Cf. R. Niebuhr, *Nature and Destiny of Man*: of these lectures, brilliant in their rhetoric but often misleading in their generalisations, it is perhaps enough to say that they only once refer, in a footnote, to J. Oman, whose *Grace and Personality* is a classic on their subject, and dismiss Dr Tennant in a single sentence as a 'Modern Pelagian' (p. 262).

[2] Of Kierkegaard the following record by his secretary is the more significant because he is apparently unaware of its symbolic meaning: Levin writes: 'He always lived on the sunny side, but shut out the sun and covered the windows with white curtains, paint or blinds'; cf. A. Dru, *Journals of Kierkegaard*, p. 563.

VI

THE INTELLECTUAL TASK: INTEGRITY

This world with all its sweep of content and of extent taxes utterance to indicate. Yet it is given us in so far to seize it, and as one coherent harmony.   SIR CHARLES SHERRINGTON, *Man on his Nature*, p. 401

1. THE NEED TO 'LEARN WHAT IS'

So from this all-too-brief survey we turn to the task that awaits us. If for convenience its three aspects are treated separately, they must plainly be undertaken simultaneously. They are in fact one and indivisible. Nevertheless

'Thy task
Is first to learn *what* is.'

And here the most serious effect of the conflict between science and religion is precisely that it destroys any possibility of an integrated *Weltanschauung*. At a time when the vast expansion of human knowledge has made specialisation inevitable and when in consequence a synoptic outlook which has surveyed the entire field, seen its various areas in true proportion, and formed a consistent picture of the whole is difficult to attain, the breakdown of the traditional Christian schema, and the failure to produce a satisfactory substitute or reinterpretation have proved disastrous not only to individual lives but to civilisation.

With the latter we have become tragically familiar; for even if the present war can only be described as a clash of rival ideologies by a process of rationalisation, yet it is manifest that the breakdown of international comity is not solely due to economic or political causes. Whereas in 1914–18 we had a conflict fought out between countries professedly Christian, and therefore not involving any fundamental differences in the way of life, in this war Christianity is repudiated and persecuted by many of the

80

combatants and it seems clear that victory will therefore have more profound consequences than in any other struggle in recent times. If Russia or Germany triumph, we may well see an organised attempt to shape human life on a pattern radically opposed to that of Christendom. If the strife is prolonged, it seems only too likely that the combatants will renounce (as we in this country are gradually renouncing) all those restraints by which generations of Christian effort have striven to humanise the bestiality of total war.

It has, of course, long been obvious to most of those who work for international co-operation that unless the universal acceptance of the externals of our culture were accompanied by a measure of agreement as to the meaning and value of existence the danger of catastrophic violence was hardly avoidable. To give to mankind in general the power to manufacture and use modern means of destruction; to encourage, as the armament manufacturers in Europe and America avowedly encouraged, the wholesale distribution of high explosives and poison gas; to foment as they have deliberately fomented the desire to test out their novelties in war; this is a sufficiently appalling story. That it was done with the definite intent to sabotage international concord in the interest of private money-making, only makes the story more devilish. But that serious students, passionately concerned for the League of Nations and for the unifying of mankind, should have argued that a civilisation whose unity was found in common systems of sanitation or transport, of industrialism and athletics, could afford to ignore fundamental differences of religion and philosophy, this was scarcely less culpable. Yet grave and learned advocates of disarmament and peace were apparently prepared to leave it entirely open whether Islam with its doctrine of the Jehad or Buddhism with its ethic of world renunciation or Christianity with the Cross as its sign were the religion of the future. Presumably they assumed that their own placid agnosticism and sweet reasonableness was the creed of all sensible men and would have power to restrain the passions and unite the tribes of mankind.

THE INTELLECTUAL TASK: INTEGRITY

Considering how inconsistent Christians have been in this matter and how readily their religious unity is abandoned at the bidding of patriotism or politics, considering also that Christianity has only too often failed to advocate or maintain peace, it may seem hard to argue that a universal Church is an essential condition of world pacification. Christianity is not in a strong position to promote unity so long as it is itself divided into sects which quarrel and denounce. But that a unifying ideal, that is a world-wide religion, of some kind is essential would be difficult to dispute. For without it and the principles of life contained in it, any consistent co-operation, any organic community of nations and races, would seem to be wholly impossible. The conditions which make for international fellowship are the same on a large scale as those which make for integration of personality in the individual. They can be more easily studied in the smaller field and in connection with the problem of education.

Until the end of the last century the medieval tradition, modified but not radically altered by the Renaissance and Reformation, by the 'New Philosophy' and by Thomas Arnold, still prevailed in the schools and universities of England. Theology, which Origen had established as Queen of all the departments of Knowledge, had abdicated from her throne when Augustine and his successors decided that secular knowledge had no relevance for the Civitas Dei. The attempt of the Scholastics to restore her had scarcely lasted for a century and had led to the futilities of Nominalism. Long before Bacon and the adoption of the inductive method the sciences had been claiming their independence. With the acceptance of the Copernican astronomy and the Newtonian physics, with the introduction of observation and experiment into methods of study, with the foundation of observatories, laboratories and museums, the traditional curriculum disclosed its absurdity; and with it fell the old limited but coherent Christian culture.

Nevertheless Christianity remained the unifying, and to some extent integrating, subject which gave a measure of consistency to the study of all other subjects and supplied the general back-

82

ground of ideas into which they were more or less fortuitously fitted. It might not be easy to reconcile some at least of the Christian tenets with the findings of scientists or historians; nor could the objectionable legends be always discreetly forgotten or adroitly allegorised. But with an amazing persistency the Christian scheme kept its place; the few who rebelled against it found themselves unable to provide satisfactory alternatives, and in many cases where their quest was honest ended by rediscovering that from which they had broken away. As compared with agnostic, Stoic or humanitarian substitutes Christianity revealed an unexpected relevance, reasonableness and power.

But as we have seen the conflict between science and religion shook, if it did not destroy, the old scheme and outlook. The newer subjects of study claimed an autonomy which at its worst recognised no obligation towards the other members of the curriculum. Departmentalism became accepted; and under the pressure of economic need specialisation at an early age began to frustrate the possibility of educating fully developed and well-proportioned men and women. The population which theoretically had access to treasures of knowledge far richer than that of any previous age ceased to possess any general culture, any common heritage, any integrating philosophy. The increase of schools and colleges seemed to synchronise with, if it could not be proved to contribute to, the increase of mental defectives and illiterates. And Dean Inge preached eugenics and the 'narrow way'.

The position may in fact have been as depressing as he suggested, but his remedies, which are those of Nazism and the master-race, have proved worse than the disease. Fortunately there were very many who were more familiar with the facts, more capable of creative effort, and more generous in their sympathies and standards. In consequence changes in the concept, scope and technique of education gained favour and were given expression not only in memoranda and recommendations but in experiment and practice. Training of the personality for world citizenship; the consequent need for a coherent presentation of the whole field of human experience and effort; the insistence upon proportion and perspective in the curriculum; the reform of methods in every department so as to stimulate intelligence and build up character; and the consequent recognition that religion starting from the child's sense of wonder and curiosity and developing its interpretative, ethical and social abilities, must take a primary place; these things became the commonplaces of educational theory and accomplished what has been called a Copernican change in regard to the whole subject. That the vested interests of denominations, parties and classes made the application of these principles difficult; that pledges were ignored, financial ramps promoted, and the whole subject let down by a succession of ministers ill-qualified and half-hearted; and that this country is still sadly far from putting its ideals into practice; these things must be admitted. Nevertheless, if the need is granted, and the way to its fulfilment in some measure agreed, the situation cannot be regarded as hopeless; and in fact in many Christian schools and colleges overseas work of first-rate quality has been done.[1]

Undoubtedly the failure to answer the intellectual question: 'What is the explanation of this queer world—of its origin, character and destiny?' is largely responsible for the delay in putting principles into practice. Unless we can give reasonable answers to the sort of questions which every intelligent boy and girl asks—the answering of which is in fact their education[2]—we cannot begin to plan our curriculum as a whole. The child when it asks, 'Where did I come from?' is not nowadays satisfied with the fable of the stork; but that fable goes nearer to an answer than much prim talk about the biology of reproduction. The child is

---

[1] Cf. the Reports on Religious Education submitted to the International Missionary Conferences at Jerusalem, 1928 and Tambaram, 1938.

[2] That science itself arises out of wonder and curiosity not out of the wish to control is well argued by J. R. Baker, *The Scientific Life*, p. 80: his attack upon J. G. Crowther, Hogben and others is vigorous and entertaining; but it is not for an outsider to assess the verdict. Both sides seem to overstate their case.

interested (however much our sceptics may dislike the fact) in metaphysics even more than in physics, in final causes rather than in temporal antecedents, in religion rather than in science. A legend may be (and by its very nature generally is) more informative than a laboratory.

That is not to justify falsehoods or to excuse the harm that Old Testament folklore when presented as history has done to religion. It is merely to underline the simple difficulties of our task. For at the moment an agreed metaphysic is far to seek. We do not easily accept the anthropomorphisms and mythologisings of our scriptural inheritance: we do not want to present as fact what the child soon discovers to be allegory or error. But the questions are asked —persistently; and to evade them is almost as damaging as to say what we know to be untrue. Poetry, imagination, symbolism are proper, indeed essential; but we want our symbols to be capable, hereafter if not now, of an interpretation that rings true; and this is not easy unless we have a consistent philosophy at the back of our replies. To provide such a philosophy is the foremost duty of those who would plan for the future, whether in the education of individuals or in the integration and development of communities.

2. THE MEANS TO KNOWLEDGE

It is a great encouragement, and an event of profound significance, that the younger scientists should not only be recognising this duty but claiming to take a direct part in its discharge. The older men —those who fill the great chairs and form the Council of the Royal Society—may declare that if a scientist utters views upon sociology or religion he does so not as a scientist but as a citizen. The memory of past controversies bids them keep science out of politics as they would keep theology out of the modern Universities. But the younger men are not satisfied to lay aside their professional training, and refuse to confine scientific enquiry within the traditional frontiers. Dr C. H. Waddington, in his Pelican book, *The Scientific Attitude*, has done good service by insisting not only that the scientist cannot ignore any human interest, but that he has a peculiar standpoint and method which if honestly applied to the whole field would yield a satisfying and complete *Weltanschauung*. With much of the book, especially the sections on the status and prospects of Christianity and on the claims made for science, it is not easy to agree: but in its main contention that we must see life whole, and that science must help us to do so it is sound and timely.[1] It would gain in value if its author had been more ready to admit that the scientific method, on the importance of which he rightly insists, is not the sole perquisite of workers in laboratories, but is in fact applied at least as rigidly and honestly in their proper fields by historians and even theologians.

If we accept the fulfilment of our intellectual task, the first step is this recognition that the method of observation, testing, induction and the formulation of hypotheses is appropriate and universal. It was not invented by Francis Bacon; nor is its use peculiar to members of the Royal Society. It is in fact what every baby does and what every teacher (and pre-eminently the founder of Christianity) encourages. Open your eyes and see; test and notice the consequences; grasp the significance of your observations; act upon it as a means to further discoveries. There is nothing highly original in such a method, nor can any of us claim a monopoly in its use.

It is indeed noteworthy that those who insist most strongly on this method in their own field of study are often lax and unintelligent in applying it to other subjects. One or two of the scientists who have tried to write histories of science or biographies of scientists have used the same care in historical research, in gathering, testing and verifying data, as they would employ in a laboratory. But the great majority seems to have no appreciation of the need for scientific accuracy in dealing with records, in sifting evidence, in checking statements, or in maintaining a sense of proportion—that is in gauging the relation of a particular fact to the general subject. When, for example, Dr Hogben refers to

[1] H. G. Wood, *Christianity and Civilisation*, pp. 21–37, makes it unnecessary to criticise in detail here.

the early history of botany he gives an account of Joachim Jung[1] whom, for some obscure reason, he regards as a botanist of importance: in this with one exception all his statements are untrue; Jung did not mark a turning-point; he did not write a herbal; he did not arrange flowering plants in assemblages; the whole story is a myth proceeding from Dr Hogben's over-fertile imagination. When a few pages later he assures us that 'the materials of Aristotle's survey of animals then known were collected during the campaigns of Philip of Macedon'[2] or that 'a few birds like the moorhen lead a wholly aquatic existence'[3] the reader will realise that the value of *Science for the Citizen* is rather that of impressionist art than of historical science. So too when Sir William Dampier, in pardonable family pride, records the achievements of his ancestor the buccaneer at greater length than he devotes to those of John Ray,[4] and when he tells the story of twentieth-century science in terms of the High Table of Trinity College, Cambridge, he invites the suggestion that he does not fully understand the application of scientific principles to the study of history. Or when the work of the eminent scientists who have written accounts of John Ray is examined, it is regrettably obvious that Edwin Lankester, editing Ray's letters and memorials for the Ray Society, produced a series of footnotes of monumental inaccuracy; that Professor Boulger, in dealing with his youth, borrowed from a gossiping paragraph in a gardening paper a story of the local squire's benefaction which a few minutes' research would have exploded; that this legend published in the *Dictionary of National Biography* has been copied without verification by most subsequent writers; and that no one of the many who have professed to give accounts of Ray has ever even looked at the Parish Register in which he and his kinsfolk are recorded.

---

[1] *Science for the Citizen*, p. 924. A different but hardly less mythical account of early botany is given by J. S. Huxley, *Evolution: the Modern Synthesis*, pp. 263–4. It is a pity that neither of these authors has studied Dr Agnes Arber's excellent *Herbals*.   [2] P. 928.
[3] P. 943.   [4] *A History of Science*, pp. 204–5 (Dampier), 200 (Ray).

THE INTELLECTUAL TASK: INTEGRITY

It is so often and (in some cases) so bluntly asserted that science has a unique regard for truth and unique methods of attaining it, and that historians, archaeologists, metaphysicians and theologians are credulous and inexact, that it is necessary to insist plainly upon facts of this kind. Instances could be multiplied almost indefinitely from the writings of many of those who pass from science to other subjects. Error is a human failing and a scientific training, though valuable, does not confer immunity against it.

The plain fact is, of course, that scientists are very much like the rest of us. They observe—as we all do in our special fields; they test their observations by experiment—as in one way or another the historian or the theologian also tests his data; they formulate hypotheses—sometimes inventing myths like that of the all-pervading material aether which dominated physics a generation ago, or of the planetary electrons whose models were so familiar till Heisenberg, or of the quantum theory which (it may be permissible to conjecture) is perhaps a mere 'interim' formula indicating incomplete analysis;[1] and to these hypotheses they attach a more than credal inerrancy—until further research knocks the bottom out of them. It is all a normal, lovable, and entirely human procedure which only invites criticism when its devotees imitate the behaviour of other priesthoods and claim a pontifical authority for their theories.

Indeed, for at least the past century the method followed in all serious academic study has been not only similar in character but applied with equal severity and impartiality. In all subjects there is a body of orthodox dogma which serves as a norm of opinion and alters more rapidly in the newer branches of study: but no scholar

[1] Cf. A. D. Ritchie on the mythologies of science: 'It would be an interesting task to examine carefully any ordinary text-book of Physics or Chemistry and try to unearth all the myths to be found embedded there—the experiments that nobody has done or could do, the sophistries that support theories, and the sheer dogmatic assertion unsupported by even a pretence of evidence. There is much there as fantastic as primitive folklore, though not so picturesque.' *Scientific Method*, p. 158. A list of some of these 'fictions' is given by Tennant, *Phil. Theology*, II, pp. 260-2.

is committed to confirming established results, or hesitates to follow the weight of evidence wherever this leads him. Theologians have in fact been even less ready to accept 'assured conclusions' than biologists or historians, perhaps because the technique appropriate to their subject is more delicate and difficult to apply than that of a laboratory, and the results are consequently less 'assured'. But those who have been trained to work in several fields will agree that though the resources available for the supply of material and the means for its sifting, testing and assaying necessarily differ, the process of intellectual research demands the same kind of equipment. Whether we are investigating the principles of inheritance, or the biography of Mendel, or the philosophy of determinism, or the doctrine of Pelagius, the enquiry will follow much the same course. Incidentally, for any adequate interpretation of the significance of heredity, all four subjects are necessary.

There are, of course, many who will raise objections.

To the claim that the study of history is only an extension of the study of evolution they will reply (in spite of Bury's famous essay) that historians deal not with the general but with the particular, not with laws but with individuals[1]—a statement less convincing since Montesquieu and least so in these days of the Marxist concept of history. In any case the difference would seem to be due to the enormous complexity of human character and behaviour as compared with that of the rest of the animal kingdom: natural history can only be studied by observation of concrete examples, and men differ far more widely than monkeys; but, as recent work has shown, a worthy study of monkeys involves the biography of individuals and the history of families and clans; as biology ceases to be concerned only with dead beasts its methods will approximate more closely to those of the historian. It is

[1] So e.g. J. L. Stocks, *Time, Cause and Eternity*, a book to which I owe much. He affirms the contrast though admitting (p. 68) that work like the *Origin of Species* is 'in substance an attempt at the reconstruction of an historical process'.

significant that when he deals with animal psychology Dr Lloyd Morgan concentrates upon 'my dog Tony'.[1]

To our claim that research in history or theology is unbiased they will protest that it depends not upon the study of facts in general but upon selection and interpretation, that is, upon prior assumptions, a hypothesis and criteria—as if there were any conceivable subject of study in which such assumptions were not essential. As Professor Ritchie puts it: 'Darwin must have had some sort of hypothesis or he would not have known what facts to examine. There were millions of facts, and he could not attend to them all.'[2] A scientist, like a theologian, collects material in order to produce not a builder's yard but a structure;[3] he must select and arrange: and if in dealing with inanimate or animal subjects there is less room for the personal equation, yet an honest student in whatever field will do his best to secure strict testing and an arrangement in accordance with the nature of the material.

Finally to our claim that all intellectual interpretation of experience can be undertaken scientifically, they will maintain that science cannot deal with what is unique since it depends upon the repetition of the event under controlled conditions, whereas history and theology are largely concerned with events like the death of Socrates or of Christ which happened once and cannot be repeated—as if science could only take account of phenomena which could be precisely reproduced in a laboratory and must ignore the appearance of a new nebula or the extinction of the last dinosaur. Particular events, even if they are what is commonly called miraculous or supernatural, can be scientifically studied by investigation of the evidence for them and of their supposed effects, by comparison with similar or analogous pheno-

---

[1] Cf. *The Animal Mind*, pp. 5–13. [2] *Scientific Method*, p. 104.
[3] 'Science is made with facts as a house is made with bricks, but an accumulation of facts is no more a science than a heap of bricks is a house.' Poincaré, *La Science et l'Hypothèse*, quoted by Tennant, *Phil. Theology*, I, p. 344.

mena, and by relation to the general body of knowledge so far as this is appropriate to them.[1]

3. THE URGENCY OF THE ISSUE

If it be objected that by stretching the term 'science' to cover such a wide field we are making it synonymous with knowledge, then it can only be said in reply that provided we recognise the need for an inclusive and coherent study of the whole universe of our experience and recognise too that the realm of weight and measurement is neither the whole nor indeed the most important sphere of human interest, the exact definition of science can be a matter of further discussion. We shall do our best to treat truth as one and ultimately indivisible; we shall insist that for mankind the studies which deal with value and meaning are more important than the physical sciences; we shall refuse to interpret reality without reference to the characteristic human experiences of wonder and worship, of penitence and dependence, of morality and religion; we shall not hesitate to use evidence from one field of study to illuminate problems in another; we shall strive towards an exact, orderly and proportionate understanding of our world in order that by knowing how things work we may gain insight into their meaning.

For it must be said plainly and strongly that the interpretation of human experience in terms which deal only with weight and measurement not only fails to describe anything except the relatively insignificant, but distorts truth by ignoring, even when it does not explicitly deny, all that is characteristically human. As Professor Ritchie puts it: 'Playing the violin can truthfully and coherently be described as rubbing the entrails of a dead sheep

[1] The argument on similar lines developed by H. Dingle, *Science and Human Experience*, pp. 94–102, that science deals only with 'common' experiences whereas 'religious experience is always an individual matter', depends upon a distinction which seems both arbitrary and artificial: it could easily be argued that much science is individual, even legendary, and much religion universal.

with the hairs of a dead horse, but the account is trivial unless the performance is very bad'[1]—and even then the description is almost wholly irrelevant to the action. Unless account is taken of meaning and purpose, any interpretation of living organisms or indeed of the world becomes misleading; for the primary characteristic of a living organism is 'the will to live' (purpose) and the very comparison of the world to a machine involves the concept of design and control. Few examples of human bewilderment are more striking than the attempts of scientists untrained in philosophy to deal with teleology. They have absorbed the beliefs that Darwin disposed of it, and that in any case its use is unscientific and proper only to those on the 'lunatic fringe'—those who are metaphysically minded or tainted with superstition. In consequence they feel committed to its rejection; and get hot and bothered by the difficulty of the task. Here, for instance, is the effort of one popular volume: 'We must give up any idea that evolution is purposeful. It is full of apparent purpose; but this is apparent only, it is not real purpose....For evolution to be purposeful, one of two things must be true. Either living things themselves must be purposive in their evolutionary changes... or else although the living animals and plants themselves betray no purpose, purpose must exist in the mind of a divine Being.... The first is that of Bergson....The second of a number of more or less modernist theologians....It is Creationism up to date.'[2] Comment is probably superfluous: the reader will note the dogmatism of the first sentences; the false antithesis of the next; the dismissal of philosophy and theology; and the sneer at a hypothesis which is at least more plausible than its alternatives.

Similarly, when Dr Julian Huxley, after rejecting all idea of 'purpose, conscious or unconscious, either on the part of the organism or of any outside power',[3] writes: 'To read *L'Evolution*

---

[1] *Natural History of Mind*, p. 49.
[2] *The Science of Life*, by H. G. Wells, J. S. Huxley and G. P. Wells, p. 386. Huxley repeats this statement in *Evolution: the Modern Synthesis*, p. 576.　　　　[3] *Evolution: the Modern Synthesis*, p. 412.

Créatrice is to realize that Bergson was a writer of great vision but with little biological understanding, a good poet but a bad scientist. To say that an adaptive trend towards a particular specialization or towards all-round biological efficiency is explained by an *élan vital* is like saying that the movement of a railway train is "explained" by an *élan locomotif* of the engine,'[1] he invites the questions whether it is good science to equate a living organism with a dead machine;[2] whether by this analogy he means to deny that an engine is the product and expression of the purposive impulse to make travel smooth, swift and reliable; and whether he supposes as he plainly suggests that a lump of metal by a series of random, partial and undesigned changes can accidentally become a locomotive.[3] Challenged by such criticism Bergson need not turn in his grave.[4]

In both these cases we are left wondering whether the authors really imagine that the universe is the result and expression of blind chance; that complicated performances like the parasitism of the cuckoo[5] or the egg-laying of the pronuba,[6] or Sir Charles Sherrington's malaria plasmodium (where the interdependence of each element in the drama upon all the rest and therefore their simultaneous appearance are essential) have arisen fortuitously or by small and random variations; or indeed that the mechanism of which they are so fond has any meaning except in terms of purpose. If they do not, then their argument is misleading: if they

[1] *Ibid.* p. 458.

[2] Descartes' contention that animals were automata was adequately answered in 1693 by Ray, *Synopsis Quadrupedum*, pp. 1–13.

[3] Cf. Sherrington, *Man on his Nature*, p. 187: 'A machine preaches to us of purpose all the time.' Nevertheless in his account of teleology and the malaria plasmodium (*ibid.* pp. 369–75) there is bewilderment and indecision.

[4] For a much wiser estimate of Bergson cf. C. C. Hurst, *The Mechanism of Creative Evolution*, pp. xiii–xiv.

[5] Smallness of egg, short period of incubation, need for much food, rapidity of growth, restlessness, sensitiveness of back, gymnastics of ejection, are the principal factors.

[6] Cf. C. Lloyd Morgan, *The Animal Mind*, pp. 50–1.

do, then let them say so frankly and stand the consequences. As it is they seem to have inherited Darwin's complete inability to think clearly on the subject rather than T. H. Huxley's admission that teleology was inescapable.

It must not be supposed that confusion of this kind is characteristic of all scientific students when they deal with questions of philosophy. Science, like religion, is at present suffering from the effects of a period of rapid transition. Many, like Dr Hogben or Dr Huxley, who are not themselves qualified in the new physics cling to ideas which are in fact out-of-date and, like their opposite numbers in theology, make up by dogmatism and jargon for what they lack in insight and appreciation. But there are others who take a radically different position. Thus, for example, while these lectures were being written Dr F. Wood Jones, Professor of Anatomy in Manchester University, published his *Design and Purpose*, in which he insists that a disastrous mistake was made when Darwin's influence was allowed to make us forget or renounce the teleology of Paley's *Natural Theology* or of William Whewell's Bridgewater Treatise. It is a time of confusion and scientists like the rest of us are bewildered and divided about its issues.

In such a time a full acknowledgement of the primary duty of seeking an integrating philosophy; a steady effort on the part of all who are qualified to help in formulating it; a readiness to co-operate with others even if they challenge ideas which we hold dear; these would seem to be our immediate contribution. Here as elsewhere it is not in mortals to command success. The seer, the prophet, the genius are gifts for which we can prepare but which we cannot command.

Thus those of us who stand by the Christian religion and find it difficult not to say 'We have got the truth; and the world will only be saved when it accepts our faith', should nevertheless be ready to welcome the efforts of scientists, reformers and philosophers even if they seem to reject and assail our convictions. At the close of a great epoch, when the vast achievement of Pheidias, Sophocles, Pericles and Thucydides had accomplished a unique

THE INTELLECTUAL TASK: INTEGRITY

advance of human achievement, there arose the strange figure of Socrates. Seemingly unimpressed by the grandeur of his country's empire he set himself to ask questions, to challenge the assumptions and puncture the pretensions of his contemporaries, to insist upon a strict investigation of the meaning of the words that they used and the ideas that they professed. In doing so he was reckoned a mere destroyer, an atheist, an anarchist. He invited and received the martyrdom which awaits the founder of a new age. But he cleared the ground; and Plato followed.

It may well be that we who similarly stand at the end of a great period of activity must be content with what Dr C. D. Broad calls critical philosophy, 'the analysis and definition of our fundamental concepts and the clear statement and resolute criticism of our fundamental beliefs'[1] and must leave speculative philosophy until this has been done. It may even be that Dr Wittgenstein and the Logical Positivists of our own day and place, who seem to many of us to have run away from the true task of philosophy and to be wasting time and thought over barren and irritating logomachies, are fulfilling the function of Socrates, are clearing away the rubbish inevitable after such great physical and intellectual changes, are sorting out the muddle which the continuance of widely divergent opinions old and new creates, and will in time prepare the way for a Plato and the emergence of a new *Weltanschauung*.

In any case we must realise that merely to restate dogmatically but perhaps in slightly rejuvenated form the old cosmologies and philosophies will only promote and prolong confusion. An honest recognition that we are living on the edge of a new age, an honest attempt to understand and sift the knowledge available, an honest sympathy with those who share our travail even if they do not share our expectation—these would seem to be the conditions with which if we desire integrity we should do our best to comply. Only it must not be forgotten that there is need for urgency.

If our present intellectual confusion were merely a matter of academic importance it might be left to the Universities to argue

[1] *Scientific Thought*, p. 18.

95 E

out at leisure. The trouble is that the interpretation of life in categories that degrade its significance carries with it fearsome consequences for mankind. There is in the first chapter of St Paul's Epistle to the Romans a passage full of meaning in this connection. Declaring that in the observation and study of the natural order men could get sufficient evidence for the fact of God, he showed that failure to interpret this evidence had led to the idolatry of explaining the whole in terms of its part (the divine in terms of a machine or a life force!), and that this had involved moral perversion and the degradation of man by wanton pride, bestial cruelty, and internecine strife. Some day we shall realise how large a responsibility our own similar idolatries must bear for the calamities that make havoc of us.

VII

THE MORAL TASK: SYMPATHY

If we would have any content in mind and spirit, we must know aright by valuing aright.

JOHN OMAN, *The Natural and the Supernatural*, p. 471

1. THE IMPORTANCE OF MEANING

To agree upon and to formulate a description of our universe, even if the task were discharged with the strictest objectivity, would inevitably involve the raising and discussion of questions not merely intellectual but ethical. Indeed, as we have seen, even scientists who repudiate teleology and wish to confine themselves to purely biological explanations cannot in fact avoid passing from the 'how' to the 'why' of action. Even to the student of animal life the significance of function and therefore to some extent its value are matters of obvious concern; and as soon as man enters into the picture his self-consciousness raises mental activity to a new level, and makes moral issues inevitable.

The Tennysonian aphorism 'Think well—Do well will follow thought' is too redolent of the Victorian optimism and of the happy Hellenic world in which sin was identified with ignorance. Nevertheless, though a satisfying philosophy will not by itself make men good, it is, as we have suggested, easily demonstrable that a bad philosophy not only prevents them from becoming good but inclines them to the appropriate evil. Those who describe the present struggle as a conflict of ideologies are affirming that there is a direct connection between the theories of the class war or the Nordic *Herrenvolk* and the behaviour of Communists and Nazis respectively. It is (from the British point of view) perhaps unfortunate that both these heresies had their origin in this country— and both from that mild and well-meaning parson, Thomas

97

THE MORAL TASK: SYMPATHY

Robert Malthus, Fellow of Jesus College, Cambridge, and author of the *Essay on the Principle of Population.*

The matter is so relevant to our subject and so liable to be conveniently forgotten that a brief comment upon it may be permissible.

When Marx and Engels, German refugees, came to England and began their propaganda, they found the economic system formulated by Ricardo upon evidence supplied by Malthus dominating political life. Malthus had affirmed that population would always increase beyond the subsistence-level and that the only remedy for the resultant starvation, famine or war was control of the birth rate. To this Ricardo added the doctrine of the Wages Fund—that there was only a fixed margin available for the maintenance of the population and that any artificial raising of wages beyond the bare minimum could only be accomplished by the starvation of those whose pittance was taken for the increase. Thus crudely stated the Iron Law made it clear that no philanthropy, no legal interference, no collective bargaining could do more than palliate and would in fact prolong and to some extent increase the hardship of the workers. *Laissez-faire* was the only policy consistent with the facts.

The Communist Manifesto was based upon a thorough-going acceptance of the Ricardian economics. It declared in effect that as no People's Charter or any other half-measures could do any good the only hope for the workers was in revolution. Capitalism meant slavery, and slavery without term or relief: capitalism must be ended. Marx spent the rest of a long life in elaborating this thesis. He had little to say, as Lenin and his followers have discovered, about the morrow of the revolution or the means by which the proletariat could modify the Malthus-Ricardo doctrine. Fortunately events and discoveries have proved that the doctrine is not wholly true; and in fact Mill, and with him orthodox economists in general, following W. T. Thornton's book *On Labour*, abandoned it in 1869.

The Nazi heresy is much more plainly derived from Malthus

98

and Darwin in its doctrine of the struggle for existence and the survival of the fittest. By the Victorians (as indeed by Darwin himself) this doctrine was interpreted in the light of their two chief qualities—their sentimentality and their utilitarianism. All life was therefore a battlefield: Nature was 'red in tooth and claw with ravin': man, if he was to escape, must become predatory in his ruthlessness or immune by his self-protection. All life was shaped to this end: the colour of the gentian, the song of the skylark, the features of the hippopotamus were conditioned by their survival value and on strictly commercial lines. The glorification of warfare as we heard it in 1914 from Treitschke and Bernhardi, and the enlistment of the total resources of every individual as we hear it on all sides to-day, are the legitimate offspring of that hypothesis.[1]

Similarly when Sir Francis Galton, following up his kinsman's view and not uninfluenced by his kinsman's pedigree, put forward the claim that genius was hereditary, he gave the support of science to the doctrine of a chosen people. The Israelites indeed saved themselves from conceit by the belief that their selection was an inscrutable act of God and by the fact that selection did not involve either prosperity in peace or victory in war. Galton's doctrine had no such safeguards. It led straight to the Jingoism of Kipling ('lesser breeds without the law'), to the snobbishness of Dean Inge (quoting his own family and himself—'W. R. I. see Who's Who'[2]—as proof of the inheritance of brains), and to the race-pride and colour-prejudice of Mr Lothrop Stoddard.[3] It is indeed amusing, if it were not also mildly disgusting, to see a number of prominent persons who have advocated almost with passion the inescapable importance of heredity now tumbling over

---

[1] Cf. V. Kellogg, *Human Life as the Biologist sees it*, pp. 51–61, quoted and commented upon by J. H. Woodger, *Biological Principles*, pp. 474–6; also B. Bavink, *The Anatomy of Modern Science*, pp. 540–61.

[2] *Outspoken Essays*, Second Series, pp. 260–1: W. Bateson protested vigorously against such claims in his 'Science and Nationality'; cf. *William Bateson, Naturalist*, pp. 367–9.

[3] *The Rising Tide of Colour* (1920), etc.

themselves to prove that there is no such thing as the Nordic race, or any reason to suppose that a German (or presumably an Englishman) has the right to bear 'the white man's burden'. That the natives of Malaya and Burma are not quite ready to accept this sudden *volte-face* as genuine, and that the educated Indian who is apt to regard Europe as the land of the *nouveaux riches* scorns it, can hardly cause surprise—particularly as Lord Vansittart has provided a convenient alternative plea that the Nordic race, if after all it does exist, is not a *Herrenvolk* but a brutish and barbarian survival.

Whether or no, in these instances or others similar to them, it is fair to blame Malthus and Ricardo or Malthus and Darwin for the heresies to which they gave rise, may be matter for argument. Our purpose in discussing them has not been to underline what has been said of the evil effects of the conflict between science and religion, but to demonstrate the importance of the moral considerations with which scientists and Christians are alike concerned. The child's questions: What is it like? and How does it work? lead on inevitably to its further questions: What does it mean? and What do I do about it? Answers are equally urgent and even more difficult to give—the difficulty being partly due to the fact that with a curious and characteristic blindness the champions both of science and of religion have until lately behaved as if conduct would remain unchanged whatever changes took place in philosophy.

2. THE SEARCH FOR MEANING

Any discussion of the moral aspect of the relationship of science to religion may best begin from the famous Romanes Lecture of T. H. Huxley[1]—which incidentally illustrates the point just mentioned. Huxley, confident that all civilised human beings accepted

---

[1] 'The ethical progress of society depends not on imitating the cosmic process, still less in running away from it, but in combating it.' *Essays*, IX, p. 33: his conclusion is restated by Dr Inge in his famous *jeu d'esprit*, 'The Idea of Progress', *Outspoken Essays*, Second Series, pp. 158–83.

an ethic derived from and similar to the Christian, and realising that Darwin's picture of evolution was hard to reconcile with such an ethic, put forward the thesis that man's task and destiny was the reversal of the cosmic process. The world was a scene of selfishness and slaughter; man with his powers of conscious control, of co-operation, of sympathy must rise in opposition to the struggle for existence and set himself to eliminate the necessity for conflict. That he overstrained the antithesis between nature and man, is in keeping with the Victorian attitude and on that account intelligible. That he did not face the question how such a process could conceivably produce a creature capable of reversing it, is less intelligible. That the general contention with its Promethean appeal to human pride has still its advocates in the generation which found it easier to believe in Jesus than to believe in God, is testimony to the continuance of Huxley's bewilderment into our own time. But plainly a position in which it is claimed that evolution discloses one set of principles at work and man's conscience, presumably the product of evolution, sets him in antagonism to them, invites scepticism. It might be consistent with the sort of dualism that underlies the religion of Zoroaster; or with the composite theories of Graeco-Roman Stoicism: but neither of these can in the last resort give any reasonable account of the world.[1] Fortunately for our chances of making sense of life few serious students either of nature or of humanity would agree with Huxley's valuation of either of them. Nature is not so immoral or man so moral as he imagined.

If we reject Huxley's position as reflecting the traditional contrast between man and the rest of creation and as untrue to the facts both of nature and of history, and look to replace it from the verdicts of more recent scientists, we find the gravest difficulty in doing so.

(1) There are as we have seen many who wash their hands of the whole business. 'It seems to be but another case of trying to

[1] Zoroastrianism posited a basic contradiction: Stoicism never reconciled the antithesis between its physics and its ethics.

have the best of two worlds to attempt to mingle Darwinism and morality'[1]—a sentence which presumably means that biology and ethics belong to different worlds. And: 'When we bear in mind the uncertainty and extreme primitiveness of our present biological knowledge...when we recall how abstract it is...the dangers of applying it to human affairs appear to be particularly serious[2]'—a much more reasonable statement. Certainly such a warning is badly needed: almost equally certainly biological data cannot in fact be ignored either by the moralist or by the citizen: we cannot just leave them aside on the ground of their incompleteness.

(2) There are others, and they are probably the large majority if not of serious thinkers at least of ordinary students, who, while not endorsing the wholesale Victorian condemnation of nature, find the evidence of fecundity and ruthlessness, cruelty and suffering in organic life incompatible with any moral interpretation of the evolutionary process. Bishop Gore was perhaps right when he maintained that the pain of the animal world was the most serious of all objections to the Christian concept of God:[3] and Bishop Barnes, surveying the course of evolution, is constrained to suggest that there is an element of sheer amorality about it, a blind proliferating and eliminating, which raises grave difficulties for any form of theism.[4] The stories of the malarial parasite or of the liver-fluke, of the nuptials of the Praying Mantis or the egg-laying of the solitary wasps are strong medicine for those who approach nature sentimentally or take an optimistic view of the goodness of the world.

(3) Others again, protesting against the sole emphasis on 'tooth and claw', have insisted upon the elements of 'mutual help,

[1] Woodger, *l.c.* p. 472.　　　　[2] *Ibid.* p. 476.
[3] *Belief in God*, pp. 156–63.
[4] *Scientific Theory and Religion*, pp. 520–3; for a drastic criticism of his view, cf. Wood Jones, *Design and Purpose*, pp. 69–72. It must of course be recognised that neither Bishop Gore nor Bishop Barnes has expert knowledge of biology or natural history.

co-operation and self-sacrifice',[1] upon evidence like that gathered together by Prince Kropotkin[2] and an interpretation like that set out by Mr Gerald Heard.[3] They point out that whatever the details of the process its result has been the evolution of humanity and of heroes, artists, thinkers and saints; that in fact survival value does not belong to the predatory or the insensitive—these like the sea-urchin or the lobster[4] are in a cul-de-sac off the main road of development—but to the adaptable and the sensitive, to those who live dangerously and gain a high vitality and quickness of response; and that though the end may not justify the means, at least a process which has left a record of enlarged functions, richer capabilities and quickened sympathies cannot be lightly condemned. *E pur si muove*: from the faint beginnings of life at the level (perhaps) of the non-filterable viruses to the manhood of Jesus is an advance stupendous in its grandeur; and if the cost in effort and agony has been terrific, that must not blind us to the scale of what has been achieved. Whatever the quality and significance of the evolutionary process, it has at least taken place and has been, in Dr Tennant's phrase,[5] 'a *praeparatio anthropologica* whether designedly or not'—a fact which involves the affirmation that the universe is so constituted as to enable the development of fullness of life. 'Of its very nature the earth brings forth fruit', said Jesus.[6]

3. THE CONDITIONS OF THE SEARCH

Such wide differences of opinion testify plainly not only to the difficulty of reaching a verdict but to the extent to which the personal equation of the juror enters into the decision. If it is true that when men presume to judge a supreme work of art they are themselves judged by it, the same principle applies even more

---

[1] Cf. E. S. Goodrich, *Living Organisms*, p. 186.
[2] Cf. *Mutual Aid: a Factor in Evolution*.
[3] *The Third Morality*.
[4] H. Bergson, *Creative Evolution* (Eng. trans.), pp. 137–9.
[5] *Philosophical Theology*, II, p. 101.   [6] Mark iv, 28.

certainly to nature. We cannot but find in nature what appeals to or shocks our own individual sensibilities. The whole is too big for our apprehension nor have we measuring-rods adequate to reckon up its worth. An estimate based upon sufficient knowledge and upon objective study is possible only to the few. But the matter is important and demands an attempt to deal with it.

First, then, if we are to discuss the significance of nature, we must try to take *all* the facts into account. To select the earthquakes or the parasites is as mistaken, and almost as irritating, as to select the sunsets and the lilies. Nature is not an arena or a picture-house where we can indulge our sadism or tickle our taste for prettiness: it is, in our present reference, a symbol and an instrument (in its widest sense and as including the manhood of Jesus, the only symbol and instrument) of reality; to enquire of it is a religious vocation, 'part of the business of a Sabbath-day' as old John Ray put it[1]. We must try to see it whole.

This involves the effort to see it objectively, to get 'inside it' so to speak by honesty of vision and imaginative insight. Until we have disciplined our anthropomorphism and learnt by patient and accurate study to apprehend something of the limits and quality of the world as it appears to other living organisms we are hardly entitled to form an opinion. And this, as we have already seen, is a difficult and largely unexplored task. It is worth adding that to interpret in terms of mechanism is as false as and more misleading than to be frankly anthropomorphic. Descartes was far less of a naturalist, far less truly scientific in this respect, than Ray.[2]

To say so much is to make it presumptuous to say more—least of all on the matter of cruelty and pain where a shallow or easy optimism is intolerable. But it is broadly true that those who have most deeply entered into a disciplined study of nature, while appreciating to the full the elements of struggle and seeming amorality have not had any sort of doubt about the worth of life.

[1] *Wisdom of God*, ed. II, p. 164.

[2] Dr J. Needham's eulogy of Descartes (*Science, Religion and Reality*, pp. 230–2) seems to me both historically and philosophically unsound.

THE MORAL TASK: SYMPATHY

Job's verdict 'Though he slay me, yet will I trust in him'[1] is not only that of the conventionally religious. It is that of J. H. Fabre, 'the Homer of the insects', who, deeply affected as he was by his studies of the ghoulish horrors of insect life, could only conclude that slayers and slain were obeying 'a kind of sovereign and exquisite sacrifice';[2] or of Richard Jefferies struggling against failure and poverty and disease and yet drawing continually from nature his passionate aspiration for 'more life'; or of George Meredith who, if no great naturalist in the scientific sense, was an accurate and sensitive observer of earth and its folk and expressed, the more significantly for that, his own estimate,

> Into the breast that gives the rose
> Shall I with shuddering fall?[3]

That such a conviction is tested almost to the breaking-point by the vast fecundity and seemingly random recklessness of tropical life and by the wastage and savagery of the struggle to which every species and individual is committed, is recognised to the full by all who have paid serious heed to the facts. But in this respect at least the students of human and of sub-human nature join hands. Generally speaking it is the sheltered and fortunate who have denounced the evil of nature, and conversely the morbid and self-pitying who have envied the care-free joyousness of birds and flowers.[4] For those of wider experience the analogy between the different levels of life seems plain.[5] The scene of the drama changes:

[1] Job xiii, 15.   [2] Cf. C. V. Legros, *Fabre, Poet of Science*, p. 235.
[3] *Ode to the Spirit of Earth in Autumn*; cf. G. M. Trevelyan, *The Poetry and Philosophy of G. M.* pp. 155, 231. Cf. Goethe's confession: 'She has brought me here: she will take me hence. I trust her. She will not hate her own handiwork,' quoted by Sherrington, *Goethe on Nature and on Science*, p. 31.
[4] For a full exposition of the connection between a man's fortunes and his valuation of nature, cf. Oman, *The Natural and the Supernatural*, pp. 406, 417, etc.
[5] There can be few so naïvely confused as Professor Joad who declares that the evil in nature seems to him to rebut theism while the evil in man necessitates it (in the *Evening Standard*, 25 Aug. 1942).

the essential elements in it are the same. To those at all familiar with evil among men, to those who have honestly faced the mystery of iniquity in so typical a manifestation as the crucifixion of Jesus, the claim that there is some special amorality in nature and therefore some ethical problem different from those which arise out of human life and conduct will seem unproven and improbable. As evolution develops, so the scope and character of the moral issues also develop: evil and good take on new meanings; life and its problems become more complex. But that the process is analogous if not homologous at all its stages is certainly the conclusion which some of us who have spent our lives upon it are constrained to reach.

Does this mean an acknowledgement that the Darwinian principle of struggle and survival, of progress by elimination, and consequently of the universal necessity of strife is true for all life? And if so, is it possible to reconcile such an acknowledgement with any theistic or at least any Christian ethic or theology? Those are questions very widely raised, even if not very intelligently answered by the many who feel vaguely but strongly that Christians in coming to terms with Darwinism have betrayed their faith.[1]

In this connection (and it is vitally important) the following points must be borne in mind:

(1) By the general assent of all competent students the fact of evolution stands firmer to-day than in Darwin's time. That creation is a continuing process, not an act once for all, is a truth which Christians must recognise—as indeed the ablest early theologians and the New Testament quite plainly do.

(2) In this process there is evidence of what is usually called a reign of law, that is the world is neither a bedlam of anarchic confusion nor a padded cell in which the effects of error are not felt.

(3) Struggle and elimination, evil (if such a moral judgement

[1] It is this belief much more than any hankering after the historicity of Genesis that keeps alive the Christian dread of science.

is permissible) and suffering, have played a large though never a sole part in the process. That this struggle is an inevitable consequence of the capacity for alternate response (that is, that it is the consequence and the condition of the autonomy of the living organism) is a legitimate hypothesis.

(4) The precise method of evolution, the extent to which other factors than those recognised by Darwinism have entered into it, is obscure: if to-day the fact of evolution is admitted, and more is known of its mechanism, there is far less assurance, indeed far more recognition of ignorance than there was fifty years ago.

(5) Although the evidence for use-inheritance is precarious and its denial widely accepted, this seems only to apply to the position as between Lamarckism and Darwinism.[1] Few biologists would deny, even if they would hesitate to assert, the possibility of direct influence of the environment or even of habit in promoting gene-change as shown, for example, by the work of Dr W. H. Thorpe on the conditioning of insects in regard to food plant or of Dr Harrison as regards larval colour.

(6) Although the causes of mutations are almost wholly unknown, the Darwinian theory that they arise fortuitously is not consistent with the sudden appearance of new types of structure and behaviour involving the simultaneous accomplishment of a number of separate but interdependent changes, nor indeed with the 'vast multitude of facts not only analogous to each other, inasmuch as they present teleological order, but interwoven in one context' which condition our every activity.[2] Here a principle regulating their co-ordination, a principle which displays the evidence of initiative and of design, seems manifestly at work.[3]

[1] It is notable that J. H. Woodger, *Biological Principles*, p. 471, refuses to discuss this antithesis on the ground that 'the whole subject at present seems to be in a chaotic condition'.

[2] G. F. Stout, *Mind and Matter*, p. 148—a very important book; and W. R. Matthews, *The Purpose of God*, pp. 106-12.

[3] Cf. also L. Henderson, *The Fitness of the Environment*, quoted by Wood Jones, *l.c.* pp. 56-8.

THE MORAL TASK: SYMPATHY

(7) Struggle sifts and fixes and educates, and in small ways develops; there is no evidence at all that it inaugurates or creates; to talk about creative strife whether in physical or psychic or religious regard is to talk nonsense. The objections to Darwinism originally raised are still valid.[1] Progress, the emergence of novelty, manifests an urge towards fuller and more complex achievement and (it seems evident) some co-ordinating 'Organiser' or holistic principle which enables simultaneity and harmonious change. To this on its physical side recent studies of the nature and influence of hormones may supply a clue: on its psychological it is postulated and studied by the important Gestalt school.[2]

(8) Further research into the mechanism of heredity may be expected to throw light on the conditions which cause natural mutations, and so to give us a real theory of the 'origin of species'. Such a theory will be tested by its ability to explain complex structures and habits which by their very nature cannot have been built up step by step.[3]

---

[1] Cf. G. C. Robson and O. W. Richards, *The Variation of Animals in Nature*, pp. 181–316, for a full résumé of all evidence on Natural Selection; they conclude: 'We do not believe that it can be disregarded as a possible factor in evolution. Nevertheless there is so little positive evidence in its favour, so much that appears to tell against it, and so much that is as yet inconclusive that we have no right to assign to it the main causative rôle in evolution.' J. S. Huxley, *Evolution: the Modern Synthesis*, though he calls himself a Darwinian, admits that selection is 'incapable by itself of causing evolutionary change', p. 29.

[2] Gestalt is defined as a 'system whose parts are dynamically connected in such a way that a change of one part results in a change of all other parts'. K. Lewin, *Principles of Topographical Psychology*, p. 218.

[3] There are still those who ignore this original and unanswered objection to Darwin's theory, cf. J. S. Huxley, *l.c.* pp. 473–5, where it is insisted (1) that large simultaneous change cannot occur, the accidental coincidence of all the necessary steps being unthinkable, (2) that a step-by-step process sifted and fixed at each stage by selection is the sole remaining alternative—and this in spite of the overwhelming objections raised to it ever since Asa Gray's time.

THE MORAL TASK: SYMPATHY

(9) It is not unreasonable to predict that any such theory will involve considerations beyond those of physics and chemistry; or to protest that so long as biology deals only with biophysics and biochemistry it does not supply the full data for an account of the living organism.

These points being duly noted do not take us very far in our quest for meaning: they may point towards a valuation clearer and more ethically significant than has been hitherto given. But the plain fact seems to be that the claim[1] that science as at present restricted can supply a basis for morality is far-fetched and unnecessary. If science be so extended as to include not only psychology and history but philosophy and religion—if, that is, it becomes synonymous with and pays full heed to all intellectual activity—such a claim would be appropriate. Under its present limitations it cannot reasonably do more than insist that all ethical theories and systems should be checked by our knowledge of biological and physiological processes. Before any such system can be accepted we ought to be able to say at the least, 'this is not inconsistent with what we know of the principles operative in evolution' and at the best, 'this helps us to understand and make sense of the course and method of the natural order'.

It is perhaps at this point legitimate to give a personal illustration. My own serious scientific studies (very limited in scope and in length) were devoted to the field of Genetics under William Bateson in the heyday of the Mendel-De Vries hypothesis. Bateson, with his brilliantly critical mind and his outspoken energy, was a drastic astringent for a budding theologian. His inaugural lecture expounding a strictly deterministic Mendelism included an assault upon all legal penal codes on the ground that ethical like athletic capacity was as much a matter of the gametes as eye

[1] Such as was propounded by Dr C. H. Waddington and discussed in the symposium *Science and Ethics*. It is right to add that in reference to the contributions of Dr Barnes and Dr Matthews to this symposium I agree with the criticisms of Dr F. Wood Jones, *Design and Purpose*, pp. 69–72.

colour or night-blindness.[1] Such teaching and a continuous interest in ornithology and entomology gave me an intense concern with the problems of the significance of nature. The familiar difficulties of reconciling any belief in God with a world of cast-iron determinism, of ruthlessness and cruelty, were my problem for many years. It was impossible to keep scientific convictions and religious experience permanently estranged. It was not easy to make any clearly consistent picture out of what was seen from the two standpoints. Nevertheless, when it became evident by experimental evidence that the claim to an absolute rigidity of inheritance was an overstatement, and when further observation helped me to enter more fully into the life of birds and beasts, I discovered that in fact the problems confronting the student of science and the student of religion were similar if not identical. The meaning and development of life, frustration and error, suffering and pain, alternate response and freedom, the individual and society, the emergence and development of a sense of value—these and above all the paradox of good and evil were common quests for me in both my main interests as they are for all normal human beings. And with the discovery that I was constantly asking the same questions in each field came quite suddenly the knowledge that in both I was being compelled to give similar answers. One looks through a stereoscope and at first sees only two blurred because separate pictures; as one alters the focus of the instrument or of one's eyes suddenly the two images come together and the whole scene stands out solid and in perspective. It would be an exaggeration to say that my sudden awareness of a stereoscopic vision of reality seen from the double standpoint of science and of religion was immediate or complete: but it came, and with it a sense of discovery and of satisfaction. The two worlds were one and the same. The process of creation, studied in its initial and physical aspect scientifically, was the same as that which reached its most revealing phase in the records of man's religious development and had its culmination in Jesus who proved to be

[1] Cf. *William Bateson, Naturalist*, p. 328, where the lecture is reprinted.

what his followers had claimed, the mystery or illuminating event by which the nature of reality was unveiled and the meaning of evolution declared. I could echo the confession which Bergson ascribes to the philosopher who pays heed to mystic experience: 'La Création lui apparaîtra comme une entreprise de Dieu pour créer des créateurs, pour s'adjoindre des êtres dignes de son amour.'[1]

4. THE RESULTS: PERSONALITY

Of the religious content of this confession we shall speak hereafter. We can conclude our survey of the moral task that awaits us by expanding and summarising the results so far reached.

If the universe of our experience is studied and described as a whole, it displays a continuous process, neither mechanical in its operation nor inevitable in its outcome, but nevertheless moving, with a vast and most impressive impulse, towards a recognisable if still remote end. In the last lecture we quoted St Paul as showing the inescapable fate of those who pervert their understanding of life by idolatry. At the conclusion of his greatest exposition of his message to them, in the eighth chapter of the same letter to the Romans, he sketches his interpretation of the significance of the world's history. It is a world in the making, incomplete, frustrate, because pregnant with creatures which can only be produced through an age-long agony of trial and error, defeat and endurance. These creatures must be free in their choices—hence the fact of evil; they must be fully sensitive to joy and therefore to pain; they must be at once mature in their individual lives and wholly co-operative with one another in the life of the community; as such they can properly be called the sons of God, the perfect objects of his love because perfectly responsive to and inspired by his Spirit. In this creative process God is himself involved, indeed embodied, at each level in the appropriate degree: he is revealed incarnate in Christ; and by a continuous incarnation brings into being and into full and free

[1] *Les Deux Sources*, p. 273 (English trans. p. 218).

communion with himself creatures worthy of his love, creatures who can be truly called his family.[1]

From the possibility of alternate response which distinguishes the living cell from the blob of colloid, through the tropisms and reflex actions of simple organisms to the beginnings of foresight and the conscious choice of means, on to the wide-ranging freedom of the higher animals and the almost limitless vagaries of the human will, and so to that perfect liberty which in voluntary acceptance of love's constraint can be described by St Augustine as the blessed necessity of not doing wrong; from the elementary use of trial and error as Dr Jennings finds it among the protozoa, through the amazing sequences of instinctive behaviour with the gradual increase of evidence of intelligent control, on to the emergence of reason and the coming of self-consciousness and the development of abstract thought; from a bare relatedness through the building-up of social and sexual life to the creative intimacies of mating, parenthood and community, through the increase of sensitiveness and range of experience and so by the discipline of suffering to the sympathy whereby fully developed individuals find themselves co-operating spontaneously and immediately in the life of the society; these are the main features of the story. Freedom, reason, love; each still incomplete, but together characteristic of the whole evolutionary process and constitutive of the fully personal life which is its goal.

[1] Rom. viii, 19–26.

VIII

THE RELIGIOUS TASK: COMMUNITY

> In presence of mystery, the spiritual attitude, if monistic, cannot as I
> think be other than mystic.... If it savour of mysticism to say that Divine
> Personality shines through the Unique Individuality of the Christ, are
> not all who subscribe to a Logos-doctrine mystics?
>
> C. LLOYD MORGAN, *Life, Mind and Spirit*, pp. 312-13

1. PERSONALITY AND RELIGIOUS EXPERIENCE

AN integrative philosophy and a fully developed personal life—
how can the attainment of these ends be promoted? What
sort of scientific and therefore realistic study can support
the belief that mankind with its animal inheritance and native
egoism is capable of such achievements? The disillusionment and
disasters of the time do not suggest a very encouraging reply.

Yet in fact these ends are involved in man's primal and characteristic equipment, in the fact of his self-consciousness and power
to apprehend his environment as over against himself, that is in
his capacity for religious experience of the kind known as mysticism.

The matter is of course difficult and deserving of far fuller study
in the light of comparative psychology than it has yet received.
But briefly stated it seems evident that the specific human quality
does not consist in man's reasoning powers—Professor Köhler's
researches have shown that the higher apes display intelligence of
a rational kind;[1] nor in his appreciation of value—for here again
a naïve delight in beauty for its own sake would seem to be implied
in the 'gardens' of the bower-birds, and the rudiments of a moral
sense in the social life of rooks or parrots. What distinguishes man
from all other organisms is what Whitehead calls his solitariness

[1] *The Mentality of Apes.*

and what Oman calls his sense of the sacred: and these two are surely the obverse and reverse of the same quality.

Whether we test it by anthropology or by introspection the primitive awareness of the universe as something other than and objective to himself, and as evoking a response at once of awe and of attraction seems to be the specifically human achievement and the origin of man's aesthetic, intellectual and moral development. For it is the stimulus of the 'mysterium tremendum et fascinans' which produces the effort to localise and symbolise it in totem and fetish, to explain it in folk-tale and myth, and to propitiate it by taboo and ritual. It is this stimulus which gives to man the integrative purpose in his individual and collective life; this which finds its fullest expression in the religious geniuses, the great saints and the great mystics.

In such a summary two points, the first and the last, need expansion.

That the holy or sacred plays a primary part in man's development is recognised by an impressive list of psychologists and philosophers. Otto's great book, *The Idea of the Holy*, which first made the claim familiar should be studied in the light of Oman's criticism of it;[1] for in conscious protest against the traditional emphasis on reason and morality Otto tended to exaggerate the element of irrational terror and so to treat man's response to the sacred as solely emotional. It is in fact the reaction of the whole self and affects the whole personality. Oman, with his strong intellectualism of outlook, himself probably represents the distinction between magic and religion as being made too early. In fact, though cultus, creed and code of ethics all contribute to the normal human response, it seems that at all levels the characteristic experience is of contact with a personal or superpersonal being, with God however crudely or abstractly interpreted. The many scientists and others who claim that the supreme reality is impersonal mainly do so because they identify personal with individual: they do not (I hope) mean that the whole is less than

[1] In *Science, Religion and Reality*, pp. 285–9.

its part, as would be the case if communion with God was analogous to a lower than personal relationship.

'Mystic' is, as the literature of the subject shows, a term liable to serious misunderstanding owing to the very powerful influence of the contemplative mysticism of Plotinus and the stress upon the Via Negativa or Way of Unknowing among the followers of 'Dionysius the Areopagite'. Mystic in the sense in which it applies to religious genius does not mean 'ecstatic', nor is mysticism equivalent to other-worldliness, quietism or non-attachment. True mysticism, as Bergson and many others have pointed out, culminates in action;[1] it is inspiration, a union with God, a possession by God, so complete as to involve a genuine identification of will; it is life at its most dynamic and effective, life spontaneous, sensitive, creative; it is that 'doing of the works' which the Fourth Evangelist joins with the Psalmist in proclaiming to be the supreme function of life.

That mysticism thus understood is the direct descendant of the primitive sense of the sacred can be shown by such an examination of the history of religion as Oman has carried out in his great book, *The Natural and the Supernatural*. He has expounded the correlation between man's relationship to the world of things and persons and his relationship to God, and has set out a classification of the great ethnic faiths in the light of the attitude towards nature which accompanies them. From the evidence underlying his work and especially perhaps from such specialised study as that of J. D. Unwin's *Sex and Culture*, it is clearly demonstrated that the levels of man's physical, cultural, moral and religious life tend to rise and fall together. This comparison, while it must not be taken as endorsing the primitive view that prosperity rewards virtue in each individual case, and is indeed compatible with the suffering and martyrdom of pioneers in every field, yet demonstrates the integral character of religion and indicates its unique importance. The life of man and of human society, despite all its inconsistencies,

[1] 'Le mysticisme complet est action', *Les Deux Sources*, p. 242 (Eng. trans. p. 193).

is a whole; and religion, far from being an irrelevance or an extra, is an accurate indication and criterion of attainment in other fields.

2. THE UNITY AND DIVERSITY OF RELIGION

It is not possible here to attempt any survey of the general history and varieties of the religious experience; but certain points must be made.

The experience itself seems to be not only common to men of all ages and races but in spite of very wide divergences in the description of it to be essentially homogeneous and easily recognisable. 'Mystics of every age are akin: there is no speech or language where their voice is not heard.'[1] By general testimony it is a conviction, an overwhelming conviction, of the oneness, permanence and universality of reality: this reality is therefore beyond the categories of space-time, is infinite and indefinable. It must be admitted that in their efforts to define it[2] many mystics have deserved the rebuke that they are talking nonsense;[3] and in all cases the descriptions are inevitably coloured by the circumstances and ideas of the individual adept. Nevertheless, it is evident that the experience preserves the twofold quality of the primitive holy: reality is both overwhelmingly other than and intimately one with the mystic;[4] it judges and abases and also welcomes and integrates; it produces a clarifying of consciousness, a sense of

[1] H. B. Workman, *Christian Thought to the Reformation*, p. 193.

[2] Of which Inge, *Christian Mysticism*, pp. 335–48, gives a long catena. His own (*l.c.* p. 5) 'the attempt to realise in thought and feeling the immanence of the temporal in the eternal, and of the eternal in the temporal' should have action added to thought and feeling.

[3] So e.g. Tennant, *Philosophical Theology*, I, p. 321. The association of mysticism with the irrational calls forth very necessary protests from thinkers like Tennant and Oman.

[4] This paradoxical form of description seems more adequate than the familiar 'transcendent' and 'immanent'—both of these terms being liable to mislead.

detachment, a dispassionate vision of himself and his world and also an assurance of intimate union with all that is, so that he is caught up into the life that permeates and sustains and gives worth to the universe.[1] Reality is thus at once static in that it is itself the unmoved mover, the peace of God, the principle of eternal life, and also dynamic, the creative energy that 'rolls through all things', the love that is not passive but active and activating, the life which, though itself beyond all relative values, is the source of all beauty and truth and goodness.

If the experience is essentially the same for all mystics, its effects, though generally similar, are always powerfully influenced by the mode of feeling, thought and action into which the experience is translated; and there has often been the risk of its distortion by the influence of the cultus, creed or code of the society to which the mystic belongs. Basically there is plain evidence that all genuine mysticism produces a characteristic change of outlook towards mankind. 'L'amour qui le consume n'est plus simplement l'amour d'un homme pour Dieu, c'est l'amour de Dieu pour tous les hommes. A travers Dieu, par Dieu, il aime toute l'humanité d'un divin amour.'[2] The two commandments which sum up the religion of the Bible, the love of God and the love of mankind are one and indivisible. 'If a man say, I love God, and hateth his brother, he is a liar'[3] and conversely 'He that abideth in love abideth in God'.[4] This effect, which is the expression of union with God and of the consequent release from self, is certainly the proof and criterion of the validity of any religious experience. Its fulfilment involves the exercise not only of the emotions but of the mind and indeed of the whole personality; and therefore at once necessitates and enables the integrative philosophy and fully developed personal relationships of which we have been speaking.

[1] It is interesting to note that Bergson (*l.c.* p. 226), describing mystic union, uses the same metaphor of iron and fire which Origen in the third century had applied to the manhood and Godhead of Jesus.

[2] Bergson, *l.c.* p. 249 (English trans. p. 199).

[3] I John iv, 20.   [4] I John iv, 16.

THE RELIGIOUS TASK: COMMUNITY

If man is true to the primary qualities of his own nature, he will discover in the religious experience the culmination of the evolutionary process and the clue to its meaning and direction. Or (to put the matter conversely lest as stated above it appear merely humanistic) God as apprehended in the experience of the mystic thus reveals himself as the source and ground of the existence of the Universe, as the reality and archetype of its structure, as the nisus which initiates and impels its development. Put theologically this means that Creation is Incarnation and Incarnation is Sanctification; that in these three modes the divine Being exists and manifests himself; and that the three are one.

To state such a claim categorically is of course to invite a variety of objections; and the principal of these must be examined. Briefly stated they are:

(1) There may be, to-day indeed there is, an increasing agreement that the religious experience as described is a primary and universal characteristic of mankind: but before taking it as a clue to man's future progress, it must be recognised that many psychologists have claimed to explain it in terms which deny to it all objective validity and reduce it, if not to pure illusion, at least to the operation of one or other psychic mechanism in the mystic. Those who plead for the scientific testing of all data of experience will welcome the study of religion by psychology: most of them will readily admit that in an element so general which involves an activity of the whole self there are necessarily points of contact with all the primary instincts, and that any one of these may give an exaggerated twist to religious experience. But the mere fact that different psychologists have ascribed religion to different and mutually incompatible mechanisms indicates, if it does not prove, that religion cannot be thus reduced. It is impossible to examine these criticisms in detail. But it may be permissible to say as a personal opinion[1] that psychologists must distinguish between the actual experience and the interpretation of it by individual

[1] I have discussed this subject at some length in *Jesus and the Gospel of Love*, especially pp. 67–77.

mystics; that much of their criticism refers to the latter not to the former; and that in almost no single case even of the very striking series discussed by Dr Leuba is his explanation sufficient to explain away the whole of the experience. Anyone who has ruthlessly examined and analysed his own fragmentary experience will be likely to affirm that although obvious and recognisable mechanisms have influenced both its secondary qualities and its interpretation, the primary datum in its givenness and immediacy, in its wholesome effects and lasting results, remains as an emergent reality which can only be accepted—which is indeed better attested and therefore more trustworthy than any other event of his life. In proportion as the man is fully developed and integrated, so will his religious experience be free from distortion.

(2) Granted again that the experience is universal and underlies the whole religious activity of mankind, can this be reconciled with the diverse and often contradictory systems in which religion is expressed? It is obvious that the mysticism of Buddhist and of Christian is accompanied by widely different philosophies, and that these sanction equally different ways of life. Yet even in this extreme case it is not less obvious that the mystics in the two religions are far less widely separated than the ordinary Asiatics and Europeans who constitute the massed adherents of Buddhism or Christianity and whose religion is a matter rather of temperament and circumstances than of direct experience. Between the adepts the difference is due to their valuation of the world of nature and the corresponding concept of reality; for these constrain the Buddhist to an ethic of world-rejection and consequently a denial of all divine immanence and an ideal of escape, and the Christian to an ethic of world-redemption, a sacramental and incarnational concept of God and an ideal of service. But despite this radical contrast it is clear that the negative mysticism of many Christians —a mysticism which stops short at ecstasy and does not issue in action—has strong affinities with Buddhism, and that Buddhists are not always content to deny the value of service and sanitation or to rank personal relationships on a lower level than contem-

plation.[1] The suggestion apparently made by Whitehead[2] that we should decide between them on a basis not of the quality but of the quantity of the evidence seems strangely unscientific. If a majority maintains that the ultimate reality is impersonal (and does not mean by this 'unindividual'), the conclusion would seem contrary to the whole course of evolution which has been from the inanimate through the living to the personal.[3] It is idolatry to interpret reality in categories less than the highest in our experience.

3. CHRIST THE CONSUMMATION

As is indicated by our examination of these objections, this universal religious experience whereby the mystic is released from self-imprisonment into communion with God and a new relationship with his fellows is not only a true development of the primitive sense of the sacred but is itself consummated in the life and work of Jesus.

It is indeed testimony to the disintegration of our intellectual life that so many who are profoundly concerned with the search for truth and the achievement of personality should ignore or lightly dismiss the founder of Christianity. That a person so sensitive and so deeply concerned as Mr Aldous Huxley, for example, should brush aside the whole fact of Christ with cheap sneers at 'his very inadequate biographers' and at three minor incidents recorded of him[4] shows a lack of perspective out of keeping with the rest of his book. 'If the life and death of Jesus, or anything approaching to the record of them, actually happened,

[1] Cf. e.g. the very influential Amida cult or the tendency to give a positive meaning to Nirvana.

[2] Cf. *Religion in the Making*, pp. 61–7—the passage though often quoted is obscure.

[3] Those who speak as e.g. A. Huxley, *Ends and Means*, p. 295, of 'impersonal consciousness' are either producing a contradiction in terms or they mean super-personal or unindividual: for a brief but sufficient comment, cf. H. H. Farmer, *Experience of God*, pp. 41–2.

[4] *Ends and Means*, p. 238.

then these are facts which any view of the world which claims to be true must take into account': that sentence of Professor Dorothy Emmet[1] is worth underlining—even if the need to do so is hardly creditable to Christian scholarship or to the Church. To discuss religion and the mystic experience without reference to the New Testament, to seek for an integrative philosophy and neglect the Christendom out of which science and its consequences have sprung, to plan for the future and ignore the Church and the resources of its life in men and women of our own day[2]—these, which are only too common an attitude, go far to explain and to perpetuate our present distresses.

From what has been said the following propositions emerge. (1) The religious experience which is essentially infinite and indescribable must be expressed by the most adequate symbol that mankind can employ: the varieties of religious experience are as we have seen largely due to the varieties of the symbols by which the experience is translated for us into terms that we can comprehend. (2) The fully developed man, not merely by his words and deeds but by the quality of his personality, is for mankind the appropriate symbol and translates the content of religious experience for us more adequately than any other medium: personality is the highest and holiest fact within our human comprehension. (3) So interpreting God the perfect man not only brings him within our range but enables us to enter into the closest communion with him; for thus the wonder and fascination of the holy is quickened into the love of person for person—a love which neither exploits nor sentimentalises—a love which when men first experienced it in and through Christ demanded a new vocabulary for its expression, and initiated a world-transforming movement by its power. (4) Men thus shown God in terms of their own humanity are not only released from self-centredness but integrated into a

---

[1] *Philosophy and Faith*, p. 102.
[2] It is perhaps permissible to draw attention here to Mr F. Brittain's memoir, *Bernard Lord Manning*, published while these lectures were being delivered and illustrating these resources.

true community: we become what we love, and finding a common loyalty and a common service with our fellows discover our organic unity with them. These four propositions are in fact fulfilled in the events recorded in the New Testament. Men found in Jesus the embodiment of their highest intuitions and aspirations. As they centred upon him their loyalty and service, they found new and full life both individually and as a society. In the dramatic event of his death and resurrection and in the sacrament in which this event was perpetuated they found a mystery which disclosed the significance of life, initiated them into a new experience of God and a new relationship with their fellows, gave them a new sensitiveness and honesty and sympathy, and welded them into a fellowship such as the world had never before seen.

To make such claims is to lay upon oneself the obligation of substantiating them. And here the considerations specified in the first of these lectures must be fulfilled.

A. Clear vision, honest study, a truly scientific testing of the data for the fact of Jesus are the first necessity. Theologians have done, as has been said, a vast deal of meticulous and elaborate research into the literature of the New Testament, into the original wording and exact exegesis of the documents, into their sources, composition, authorship and reliability, into the history and circumstances which preceded, accompanied and followed them. It is safe to say that nothing in all human experience has been investigated so thoroughly or with such a concentration of effort or by so wide a variety of persons. That process must continue. For Professor Emmet is surely right when she says: 'It makes all the difference whether we can believe that we are dealing with people's attempts to interpret a real happening or whether we have a mere creation of what M. Bergson calls "la fonction fabulatrice".'[1]

This is primarily the historian's business, though he will need the help of a wide range of specialists on the technical side, and of people of insight, imagination and scientific knowledge, to give

[1] *Philosophy and Faith*, p. 102.

proportion and exactitude to his work. To share in that work to
the full extent of his power should be a normal obligation for us
all, if like the Greeks who came to Philip 'we would see Jesus'.

B. We study that we may interpret. Given the facts scrutinised,
sifted, attested, what is their significance? Here is a task equally
important and perhaps more difficult. The data for the first task
are relatively few; the data for the second involve the whole range
of human life if as we have claimed the fact of Jesus is one which
all must take into account. Moreover, it is a fact which has been
subject to such interpretation from the very first. In order to see
it clearly we had to disentangle it from the inferences, interpola-
tions and accretions which even at the beginning were involved
with it. In order to discover its full meaning we must examine
and assess the value of the various theologies and philosophies to
which it has given rise: the experience embodied in Christian
liturgies and devotions, art and music; the history of Christendom, the lives of the saints, the expansion and the varieties of
the Church; the comparison and mutual interaction between
Christianity and other religions; the relationship of Christianity
to secular history and thought, and especially to the findings of
modern science. A Christian *Weltanschauung* involves a Christ-
centred interpretation of the universe; and the clue to this will be
found in a recovery and exploration of the Church's faith in the
Holy Spirit.

C. To study and interpret will necessarily (if we are honest)
involve the effort to express our convictions in our way of life.
It is here that the claim presses hardest upon us. So long as
mankind sees in the Christian community a character hardly
different from its own, a disunion and sectarian bitterness ill-
excused by insistence upon the vital importance of the points at
issue and in flat denial of the primary obligation of love to God
and neighbour, an obscurantism which identifies faith with credulity and boasts of believing what it knows to be untrue, an
acquiescence in moral and social evils which science can remedy
and which only ignorance or prejudice can defend—so long is the

possibility of a general regard for the fact of Jesus inevitably frustrated. It was because in the early days the pagan world was compelled to say with honest admiration 'See how these Christians love one another' that the faith won its way with such amazing speed and power. Until the modern note of irony is wholly removed from those words, Christendom will have failed. A world broken by its lack of integrity and of sympathy, a world eager for community, need not surely appeal in vain to the ambassadors of Christ to come over and help us.

That appeal, which when first made was the prelude to the evangelising of Europe, seems to have initiated the partnership between St Paul the Jewish scholar and St Luke the Greek doctor. It is no doubt fanciful to see in the words 'straightway we endeavoured to go into Macedonia' a prophetic foretaste of the form in which the gospel made its way. But it was the precise combination of Hebrew religion and Greek science which gave to early Christianity its compelling power and won Europe for the faith.

We have been concerned in these lectures with the quarrel between religion and science, with its tragic consequences in the bewilderment, corruption and destruction of mankind. We are living in times when it is hard to keep far from despair or at least from overwhelming depression of soul. For very many of us the sheer pain of present and future has brought with it a sense of dereliction, a sense not merely of our own helplessness and sin, but that God has cast us off. For such there is a passage in Dr A. Nairne's book, *The Faith of the Old Testament*, which is singularly appropriate in its references to their state and to our subject.[1] He writes of Job: 'The climax of the drama is the answer of the Lord out of the whirlwind, and that is not an answer to the problem of suffering, but to Job's sense of desolation. Job's suffering was not what his friends thought. It was a prefiguring of our Lord's cry on the cross "Eloi, Eloi, lama sabachthani".

[1] This passage was sent to me by the Rev. H. St J. Hart, Dean of Queens' College, Cambridge, while these lectures were being written.

He was ready to endure all loss and pain if only his communion which seemed broken might be renewed. The book ends with no solution of the problem, but it does end with satisfaction to Job's yearning for communion. That satisfaction came partly through his eyes being opened to the largeness of God's presence. While God was caring for all nature, the obscuration of his presence in one man's heart did not mean that he was really absent from that man. Let him think largely, naturally, divinely of God, and he would not be impatient. Job learned what Hort learned from nature: "As we can seldom bathe ourselves in the freshness of living things, without coming forth with purified and brightened hearts, even such let us believe may be the effect of the truth of nature on our thoughts of God Himself" (*The Way, the Truth and the Life*, p. 84). This is an aspect of the book of Job of peculiar interest in these days of ours. Our theology has perhaps been too long confined to the isolated aspirations of mankind. A while ago man was chiefly contemplated as being very far gone from righteousness and in need of rescue. Of late his state is beginning to be rather glorified than deplored, as though he were already lifted high in Christ. A truth indeed, but a perilous one; and difficulties increase as schemes are reformulated. And many of these difficulties arise from the larger and exacter science of nature which is characteristic of the age. Yet that new science brings new reverence, new possibilities of faith with it. The hour is coming when we shall invigorate theology by recovering the Alexandrine doctrine of Christ as the Word of God, as being not merely the Saviour of men but the Redeemer of the whole creation which has been created through him. When that is done, the mystery of the person of Christ will become more intelligible again and more dominant in the general mind, and salvation will again be sought and accepted by all men as it includes men and nature in its vaster sweep.'[1]

[1] *L.c.* pp. 118–19.

125